Water Utility Benchmarking
Measurement, Methodologies, and Performance Incentives

Water Utility Benchmarking
Measurement, Methodologies, and Performance Incentives

Sanford Berg
Director of Water Studies, Public Utility Research Center,
Warrington College of Business Administration, University of Florida

With the assistance of Maria Luisa Corton, Chen Lin, and
Guillermo Sabbioni, Research Associates, Public Utility Research Center
and
Liangliang Jiang, Aaron Jones, Brandon Phibbs, and Michelle Phillips,
Research Assistants, Public Utility Research Center

IWA Publishing
London • New York

Published by **IWA Publishing**
Alliance House
12 Caxton Street
London SW1H 0QS, UK
Telephone: +44 (0)20 7654 5500
Fax: +44 (0)20 7654 5555
Email: publications@iwap.co.uk
Web: www.iwapublishing.com

First published 2010

Typeset in India by OKS Prepress Services.

British Library Cataloguing in Publication Data
A CIP catalogue record for this book is available from the British Library

Library of Congress Cataloging-in-Publication Data
A catalog record for this book is available from the Library of Congress

ISBN 10: 1843392720
ISBN 13: 9781843392729

Table of Contents

Appendices:

Preface

This volume draws upon material prepared under contract to the World Bank and the Water and Sanitation Program, through IBNET: "Survey of Benchmarking Methodologies" (2006). The report has been revised and updated, building upon a Public Utility Research Center (PURC) study funded by the Inter-American Development Bank: "Benchmarking Water Utilities in Central America" (2007). In addition, the Public Utility Research Center, the Warrington College of Business Administration, and the University of Florida Water Institute have provided support for research leading to this book. We are grateful to those who have provided input into this multi-year initiative. The volume consolidates PURC's research in water utility benchmarking and extends the International Water Association's set of publications on this topic. The purpose of this book is to provide an overview of the strengths and limitations of different methodologies for making performance comparisons over time and across water utilities (metric benchmarking). In addition, it identifies ways to determine the robustness of performance rankings. While many of the lessons are especially applicable to developing countries, the methodologies introduced and the incentives described here are relevant for decision-makers in OECD nations as well.

Benchmarking is essential for those developing and implementing water policy. If decision-makers do not know where they have been or where they are, it would seem to be impossible to set reasonable targets for future performance. Information on water/sewerage system (WSS) operations, investments, and outputs is essential for good management and governmental oversight. This book is designed to help decision-makers to identify the data required for performance comparisons over time and across water utilities, to understand the strengths and limitations of alternative benchmarking methodologies, and to perform (or commission) benchmarking studies.

The target audience for this volume is senior staff of regulatory agencies, professionals in related government agencies, water utility managers, researchers, consultants, lawyers, civil servants, and staff of international financial institutions. The volume provides a generalist with an overview of the different methodologies used to compare performance of WSS operations. The book does not attempt to provide insights on how specific production processes might be improved (process benchmarking). Rather, the focus is on the implications for incentives—established by those regulating utilities and by utilities for strong managerial performance.

Technical material is placed in Appendices. Data definitions are presented in Appendix 1. Specialists should benefit from Appendix 2, an Annotated Bibliography which surveys eight illustrative benchmarking studies. Appendix 3 reviews some technical material, but those conducting studies will find more comprehensive discussions of the metric methodologies in the scholarly literature. In addition, Appendix 4 briefly surveys current benchmarking activities in Latin America, Asia, Africa, and OECD nations.

The performance comparisons draw upon actual cases and examples to illustrate applications of the methodologies. The volume underscores the importance of obtaining consistent data for a number of years to facilitate quantitative analyses. Five basic approaches to benchmarking characterize current studies:

- Core Indicators and a Summary or Overall Performance Indicator (partial metric method)
- Performance Scores based on Production or Cost Estimates ("total" methods)
- Performance Relative to a Model Company (engineering approach)
- Process Benchmarking
- Customer Survey Benchmarking

The focus of this study is on the first two approaches: the creation of partial and total metrics. Such metrics do not identify specific stages of the process that fall short of best practice, nor do the results of metric benchmarking studies show managers how to implement improvements. The purpose of this book is to demonstrate how analysts can *measure* water utility *operations* (inputs and outputs) to *perform company comparisons* in the context of *infrastructure reform*. The focus will be on Performance Scores based on comprehensive production and cost studies; however, no single methodology captures all the elements that are relevant for ranking utilities (or nations) in terms of water sector performance.

Recommended as a companion volume is the following book: *Performance Assessment of Urban Infrastructure Services: Drinking water, wastewater, and solid waste* (2008) edited by Enrique Cabrera Jr. and Miguel Angel Pardo

(IWA Publishing). Its chapters are from the Proceedings of the Pi08 Conference in Valencia, Spain. The event was the final stage of the COST C18 Action, a network of experts funded by the EU—bringing together practitioners, consultants, and academics from many nations. The numerous case studies and analytical reports represent a wide variety of approaches to performance assessment. However, only a few of the national studies utilized the metric benchmarking techniques described here. So there is certainly a need for professionals concerned with water utility performance assessment to become more familiar with economic approaches to the analysis of cost and production.

Another useful volume is *Benchmarking Water Services: Guiding Water Utilities to Excellence* (2010). This IWA Manual of Best Practice is edited by Enrique Cabrera Jr., Peter Dane, Scott Haskins, and Heimo Theuretzbacher-Fritz. It draws heavily from the European experience and engineering techniques as applied to benchmarking. The present book differs from the IWA manual by focusing on cost and production function approaches utilized by economists in analyzing water utility performance. In addition, here, benchmarking is taken to be equivalent to performance assessment.

The primary author appreciates all those individuals who have assisted in the development of the material presented here. In addition, he is grateful to organizations that have supported this line of investigation over the past five years.

Dedication: This volume is dedicated to those who are attempting to bring quantitative methods to bear on the important issue of incentives for improving water utility performance: technical researchers, regulators, policy-makers, and managers—all those seeking safe (and affordable) water for every person on our planet.

Acknowledgements

The PURC Team would like to thank Caroline van den Berg (World Bank), Alejo Molinari (ADERASA Benchmarking Task Force), Silver Mugisha (National Water and Sewerage Company, Uganda), Chunrong Ai (University of Florida), Lynne Holt (PURC), Rui Marques (Technical University of Lisbon), Federico Basañes (IADB), and Ruben Avendaño (IADB) for suggestions and assistance at different stages of this project. The author is grateful to Lisa Jelks and Patricia Casey (copy editors) for their careful work on this volume at different stages of its development. The primary author accepts full responsibility for remaining errors of omission or commission.

Acronyms and Abbreviations

AC	Average Cost
ADERASA	Water Regulators of the Americas (Asociación de Entes Reguladores de Agua Potable y Saneamiento de las Américas)
AWWA	American Water Works Association
CAPEX	Capital Expenditures
CAPM	Capital Asset Pricing Model
COLS	Corrected Ordinary Least Squares
CRS	Constant Returns to Scale
DEA	Data Envelopment Analysis
EBIT	Earnings Before Interest and Tax
ECA	Europe and Central Asia
EUWI	European Union Water Initiative
GDP	Gross Domestic Product
IBNET	Water and Sanitation International Benchmarking Network
LRAC	Long Run Average Cost
MC	Marginal Cost
OECD	Organization for Economic Co-operation and Development
OFWAT	England and Wales regulator (Water Services Regulation Authority)
OLS	Ordinary Least Squares
OPEX	Operating Expenses
OPI	Overall Performance Indicator
PI	Partial Indicator
SAAWU	South African Association of Water Utilities
SEAWUN	Group of water associations in South East Asia
SFA	Stochastic Frontier Analysis
SNIS	Brazil National Sanitation Information System
SUNASS	Peruvian Water Regulator (Superintendencia Nacional de Servicios de Saneamiento)
TC	Total Cost

TE	Technical Efficiency
TFPC	Total Factor Productivity Change
VRS	Variable Returns to Scale
WACC	Weighted Average Cost of Capital
WHO	World Health Organization
WSS	Water Sewerage System
WUP	Water Utility Partnership for Capacity Building (Africa)

Chapter 1 Introduction

Benchmarking is essential for those developing and implementing water policy. If decision-makers do not know where they have been or where they are, it would seem to be impossible to set reasonable targets for future performance. After all, operators can only manage what they measure. Similarly, municipal, state, and national regulators (those setting policy priorities), need to understand how operators perform relative to comparable organizations. Thus, information on water/sewerage system (WSS) operations, investments, and outputs is essential for good management and effective oversight. This volume is designed to help decision-makers (1) identify the data required for performance comparisons over time and across water utilities, (2) understand the strengths and limitations of alternative benchmarking methodologies, and (3) perform (or hire consultants for) metric benchmarking studies.

Metric benchmarking quantifies the relative performance of organizations or divisions, controlling for external conditions. This book surveys how analysts can measure water utility operations (inputs and outputs) to perform company comparisons in the context of infrastructure reform. Using well-established empirical procedures, analysts can measure performance and identify performance gaps. The tools are important for documenting past performance, establishing baselines for gauging productivity improvements, and making comparisons across service providers. Performance scores are like yardsticks that compare how firms measure up to the best performing firms: they provide rankings that can inform policymakers, those providing investment funds (multilateral organizations and private investors), and customers.

Yardstick comparisons have many audiences, each with different degrees of expertise and interest when it comes to evaluating water utilities, yet each has an expectation shared by the others: rankings should reflect reality. Results that are highly sensitive to model specification or the inclusion (or exclusion) of particular variables and data points will not be credible. If the criterion of consistency is not met, these groups cannot be confident that the relative performance indicators are meaningful. Thus, if alternative methodologies do

 Water Utility Benchmarking: Measurement, Methodologies, and Performance Incentives. By Sanford V Berg. ISBN: 9781843392729. Published by IWA Publishing, London, UK.

not yield broadly similar rankings, analysts should be able to explain the discrepancies and use the resulting rankings with caution.

Data: The first step in benchmarking involves collecting information on water/sewerage system (WSS) operations, network capacity, financial flows, and outputs. Consistent data that facilitate comparisons are essential for good management and for public policy oversight; such data are becoming available via the *Water & Sanitation International Benchmarking Network* (IBNET, funded by the UK Department for International Development and the World Bank). Some regional water operator associations are also beginning to publish data on participating organizations. For example, the UN-HABITAT and the International Water Association are collaborating to strengthen Water Operators Partnership activities (including data collection) in various regions of the world (See Appendix 4). In addition, to promote transparency, a number of regulatory agencies are making data available on the financial and operating performance of utilities. Nevertheless, there is much to be done to improve the collection, reporting, auditing, and evaluation of data.

Models and methodologies: When conducting benchmarking analyses, water professionals must understand the strengths and limitations of different metric methodologies. This book is directed to regulators and utility managers to help them make performance comparisons over time, across water utilities, and across countries. Although consultants and academic researchers have published over one hundred empirical studies of water utilities (eight of which are summarized in Appendix 2)[1], few regulators and companies are using metric benchmarking on a regular basis. Perhaps they are concerned that statistical and frontier studies are not robust. This study underscores the need for sensitivity tests so that analysts can be confident in the rankings that emerge from data analysis.

Interpreting studies: The application of more sophisticated quantitative tools is necessary (but not sufficient) for promoting policies that can improve company (and sector) performance. The introduction of greater rigor allows stakeholders to quantify utility progress towards meeting policy objectives, helps water specialists to identify high performing utilities (whose managerial and engineering processes might be adopted by others), and enables regulators to fine-tune targets and incentives for utilities. However, the skills and resources

---

[1] See Malcom Abbott and Bruce Cohen (forthcoming). "Productivity and Efficiency in the Water Industry," *Utilities Policy*. They survey sixty-seven studies. Another survey article is by M. Walter, *et al.* (forthcoming). "Quo vadis efficiency analysis of water distribution? A comparative literature review," *Utilities Policy*.

required to conduct (or monitor) benchmarking studies are in short supply. In addition, communicating the results to various constituencies requires great care. This book is designed to bridge the gap between technical researchers and those practitioners currently conducting (or contemplating) studies for government agencies and water utilities.

If sector performance were meeting global expectations, new water initiatives would not be a high international priority. However, to even come close to achieving the Millennium Development Goals for Water, benchmarking will need to be used as a key tool to improve service quality, expand networks, and optimize operations. Although both regulators and managers are aware of benchmarking techniques, they sometimes lack the professional staff able to conduct analyses. Ideally, the water utility regulator (at the national, state, or municipal level) would review studies and create performance incentives to achieve policy objectives. Incentive regulation works: studies have shown that after controlling for physical, social, and environmental factors, utilities with clear incentive structures will improve efficiency over time, whereas those without targets and incentives will (in general) show falling efficiency ratings.[2] Without confidence in the measurements, however, those responsible for creating incentives will not risk their credibility by instituting rewards or applying penalties. Regulators will be unwilling to apply incentives based on performance unless they are very confident that the rankings can survive challenges. Furthermore, in some cases regulators may wish to avoid the political pressure generated when poorly performing utilities are singled out. "Knowledge is power," and providing information to stakeholders disturbs the *status quo*.

Many groups are potentially interested in comparisons of utility performance; they are interested in the determinants of cost that are beyond the control of management. Benchmarking specialists produce and critique studies that utilize various methodologies. This group is in a good position to validate or discredit performance comparisons. The press and other news groups filter and highlight reports, using executive summaries and interviews. This group will ferret out disagreements because it is in the interest of affected parties to dispute low rankings. The general public tries to understand the implications of rankings for judgments about water sector performance. However, citizens are not

---

[2] See Kristof De Witte and Rui C. Marques (forthcoming) "Designing Performance Incentives: an International Benchmark Study in the Water Sector," *Central European Journal of Operations Research.*

well-positioned to evaluate conflicting claims, which puts a heavy burden on those producing utility performance ratings.

Policymakers from the legislative and executive branches of government are also important consumers of information. National policymakers (elected representatives and appointed officials) react to and utilize technical studies in setting priorities and interacting with international organizations. To some extent, the absence of benchmarking information takes pressures off policymakers because citizens are unaware of performance trends and the extent to which utilities might fall short of best practice. The problems are particularly severe in developing nations. Since public investments in water systems mean less funding is available for hospitals, schools, and other social infrastructure, citizens want to be sure that water utilities are performing well. Otherwise, policymakers can posture, utilities can pretend to supply water of high quality, and consumers can pretend to pay (since revenue in many systems does not cover reasonable operating costs). The outcome damages all three groups. Without information there is no catalyst for reform. At the same time, technical studies are not panaceas for improving sector performance. Expertise is necessary but not sufficient for implementing incentives based on relative performance.

Winston Churchill said that experts should be "on tap, not on top."

The message is clear: water is a sensitive sector requiring those developing and implementing policy to balance many objectives (including affordability and network expansion, for example). Studies are inputs into decision-making. However, stakeholders will shape the way policies are implemented. Quantitative methods represent one tool available to those promoting evidence-based policies. Thus, sustainable decisions require that regulators think politically without being political.

1.1 Basic Definitions

Before summarizing five widely-used methodologies, it is useful to introduce a few definitions:

Productivity considers the link between inputs and an organization's outputs. The ratio *outputs/inputs* serves as an indicator of current performance. The business press will often use a *partial indicator* when discussing productivity; thus, we see output per worker or connections per worker used to make comparisons over time or across firms (or sectors). Although output per worker is a measure of labor productivity, in isolation it is somewhat uninformative. The use of output/labor trends over time or comparisons across firms can yield distorted results, since such indicators do not capture the roles of other inputs (or the outsourcing of particular activities). When a utility produces *multiple* outputs and uses *many* inputs, we encounter the problem of how to aggregate the components. How do we appropriately weight output quantities and input quantities to obtain an overall index? Economists use the technical term *total factor productivity* defined as the ratio of outputs to inputs: specialists have developed approaches to resolving related index number issues related to weights.[3]

Efficiency is related to productivity, but it involves establishing a standard and determining how close the firm comes to meeting that standard: how far is the utility from "efficient practice"? Basically, the question is how near the utility is to its production frontier, given its inputs. If other (comparable) firms produce more output when using the same input levels, the utility in question is relatively inefficient. Of course, no two firms are exactly alike, but information on the inputs and outputs of the best performing firms can be can be used to establish a "frontier". Efficiency can be further broken down into components: *engineering efficiency* (which does not consider input prices) and *allocative efficiency* (which checks whether costs are being minimized for producing a particular output level—sometimes labeled *production efficiency*).[4] Economists have used *production functions* to identify relationships between inputs and outputs and they use *cost functions* to show how input prices and output levels determine costs. In addition, these models can evaluate outcomes associated with both best practice and average practice.

Effectiveness refers to the extent to which a utility achieves stated objectives. If the objectives are not quantifiable, then success cannot be measured. If the goals are unrealistic, the targets are meaningless. If the goals are easily achieved, they are unnecessary (and their achievement should not result in rewards for managers and staff). Benchmarking via yardstick comparisons is one way to measure outcomes, to establish achievable targets (since relative performance

---

[3] Market prices for the outputs and the inputs are typically used to aggregate the elements of the numerator and the denominator, but analysts might use "shadow prices" if the market prices do not reflect the underlying economic values of outputs and inputs.

[4] An even more comprehensive definition of allocative efficiency involves a firm producing the right output mix and pricing the outputs at marginal cost. In addition, *dynamic efficiency* considers whether the firm's costs are at the right levels over time—reflecting research and development and the adoption of innovations. For our purposes, we will focus on production efficiency and cost minimization as our measures of efficiency.

provides evidence regarding what is possible), and to structure sound incentives for improving performance.

Later sections will examine data issues and model specification problems. For example, service quality can be viewed as an output that requires additional inputs (and involves higher costs). For now, it is enough to note that productivity measures can be used to rank firms or to establish trends over time. Efficiency scores can be used to indicate how close a firm is achieving what other firms have been able to achieve from their resources. Frontier analyses use "best practice" or best performance as the standard, but some analysts apply methodologies that utilize "average" performance as the standard (or benchmark) when evaluating managerial performance.

1.2 Five Methodologies

If a physician were to evaluate the health of another person, this medical specialist could obtain a lot of information from just four core indicators: weight, height, blood pressure and temperature. Of course, much more information would be needed for diagnostic purposes, but trends in such health indicators provide a starting point. More comprehensive tests could include blood work, a MRI, and perhaps even genetic testing. Evaluating the health of an individual or a population is a complex process. In recognition of the wide range of issues that might be addressed when evaluating water utility performance (as in evaluating a patient's health), analysts have developed a variety of methodologies for addressing specific issues:

- Core Indicators (partial metric method),
- Performance Scores based on Production or Cost Estimates ("total" methods),
- Performance Relative to a Model Company (engineering approach),
- Process Benchmarking (involving detailed analysis of operating characteristics), and
- Customer Survey Benchmarking (identifying customer perceptions).

These techniques are briefly described below; this volume focuses on the *second* methodology.

Core Overall Performance Indicators include a number of *Specific Core Indices*, such as volume billed (output) per worker, connections per worker, quality of service (continuity, water quality, complaints), unaccounted for water, coverage, and key financial data (operating expenses relative to total revenues, collections). Usually these indicators are presented in ratio form to control for the scale of operations.[5] These *partial measures* are generally available for well-managed companies, and provide the simplest way to perform comparisons: trends direct attention to potential problem areas. Policymakers often combine the specific core indices to create an *Overall Performance Indicator (OPI)*, generally using a weighted average of core indices. Thus, an OPI provides a summary index that can be used to communicate relative performance to a wide audience. Although its components are easily understandable, in practice the weights used to compute the OPI are not determined through a process that prioritizes the different indicators. For example, the OPI used by SUNASS (the Peruvian water regulator) is the un-weighted sum of nine specific indices. In addition, many factors will affect the specific indices, including population density, ability to pay (income levels), topography, and distance from bulk water sources. Finally, an OPI fails to account for the relationships among the different factors. A firm that performs well on one measure may do poorly on another; similarly, one company doing reasonably well on all measures may (or may not) be viewed as the "most efficient" company.

Performance Scores based on Production or Cost Estimates are used to identify the best performers and the weakest performers in a group of utilities. The metric approach allows quantitative measurement of relative performance (cost efficiency, technical/engineering efficiency, scale efficiency, allocative/production efficiency, and efficiency change). Performance can be compared with other utilities at a point of time and over time, using statistical and/or nonparametric frontier methods (to be discussed later). Analysts apply these quantitative techniques to determine relationships among variables: for example, utilities that produce far less output than other utilities (who are using the same input levels) are deemed to be relatively inefficient. Similarly, a utility might have much higher costs than expected (based on observations of others producing the same output level but having lower costs). A finding of excessively high costs would trigger more in-depth studies to determine the source of such poor performance—which may be beyond management's control. Thus, performance scores and relative rankings identify under-performing and high-performing utilities. The burden of proof for explaining a low ranking can be placed on a utility's management team, which may have

---

[5] Helena Alegre, Wolfram Hirnir, Jamie Melo Baptista and Renato Parena (2000). *Performance Indicators for Water Supply Services*, International Water Association, IWA Publishing, xiii–146. Also, see the volume by Rafaela Matos, Patricia Duarte, Adriana Cardoso, Andreas Shultz, Richard Ashley, and Alejo Molinari (2003) *Performance Indicators for Wastewater Services,* IWA Publishing.

a reasonable explanation for the outcome, like a terrible storm that devastated production facilities.

Rankings can be based on the analysis of production patterns and/or cost structures. Production function studies (requiring data on inputs and outputs) show how inputs affect utility outputs (such as volume of water delivered, number of customers receiving service, and service quality). Similarly, cost functions show how output levels and input prices affect costs; such models have heavy data collection and analysis requirements. One advantage of cost models is the ability to analyze components of total cost; for example, OFWAT (water regulator for England and Wales) has examined how different types of operating expenses depend on various cost-drivers, such as length of pipe, volume of water delivered, and customer density. Using data from a group of utilities at a point in time (or over a long time period) allows analysts to incorporate cost-drivers beyond management's control (such as population density or topology). In both types of studies, estimated parameters can give an indication of economies of scale and/or economies derived from the joint supply of water service and wastewater collection and treatment. Studies have also examined the relative performance of privately-owned and publicly-owned water utilities. Data availability and the issue under investigation dictate whether production or cost functions are utilized and influence the choice of analytic technique (statistical estimation or data envelopment analysis—discussed more fully in later sections).

Engineering/Model Company approach has been used to establish baseline performance in a number of countries, including Chile and Argentina. This methodology requires the development of an optimized economic and engineering model, creating an idealized benchmark specific to each utility—incorporating the topology, demand patterns, and population density of the service territory. The use of an "artificial" firm that has optimized its network design and minimizes its operating costs can provide insight into what is possible if a firm is starting as a Greenfield Project. As with any methodology, this approach also has its limitations. The engineering models that support it can be very complicated, and the structure of the underlying production relationships can be obscured through a set of assumed coefficients used in the optimization process. Furthermore, it is unclear how relevant engineering models are for evaluating systems that have evolved over time (with erratic infusions of funds). Nevertheless, the approach is used in some settings.

Process Benchmarking focuses on individual production processes in the vertical production chain. One advantage of this approach is the ability to identify specific stages of the production process that warrant attention. For example, obtaining finished drinking water involves the a number of steps including pumping up, intake, transport, clarification and filtration of groundwater as well as the purification and treatment of raw surface water. Detailed examination of production facilities and their operations would be the starting point for process benchmarking. Similar studies would be performed for distribution processes (network design, pipeline construction and maintenance), sales processes (including meter reading, data processing, billing, collections, and customer relations), and general processes (like planning, staff recruitment and retention, and public relations). Many studies have focused on process benchmarking as a mechanism for identifying potential benchmarking partners, preparing for and undertaking benchmarking visits, and implementing best practices.[6] Thus, water utility managers recognize that information sharing is a significant performance driver across companies. However, detailed process studies are less useful for ranking utility performance across operating systems. From the standpoint of public policy, there must be a clear delineation of utility obligations and regulatory responsibilities so that process benchmarking activity does not involve undue external interference with managerial decision-making.

Customer Survey Benchmarking focuses on the perceptions of customers as a key element for performance evaluation. Unlike the other approaches, this technique can shed light on consumer concerns, reflected in the number (and type) of complaints or captured in customer surveys. Customer perceptions regarding service quality are central to evaluating water utility performance. One widely-used model[7] identifies five dimensions of service quality as perceived by customers: external characteristics (tidy workplace, employee appearances), reliability (meeting deadlines, consistency in interactions), responsiveness (providing service promptly), consideration (personnel who are courteous, friendly, and helpful), and empathy (giving individual care and attention). Surveys can reveal performance gaps and identify areas of concern. Disaggregating complaints by type of customer, location, and type of complaint can help managers identify problem areas. In addition, trends over time can be used by regulators and policy-makers to evaluate utility performance.

While customer survey benchmarking enables utility managers to attain information of customers' feelings from a large sample, some responses reflect emotional attitudes that may not capture verifiable features of service delivery.

---

[6] Mats Larsson, Renato Parena, Ed Smeets and Ingrid Troquet, (2002). *Process Benchmarking in the Water Industry: Towards a Worldwide Approach*, International Water Association.

[7] Parasuraman, Zeithaml and Berry (1985) "A Conceptual Model of Service Quality and its implications for future research," Journal of Marketing 49 (4), Fall, 41–50.

Utility rankings based on survey scores might simply reflect customer sentiments rather than technical features of service quality. Of course, such sentiments warrant attention: whether correct or not, they reflect perceptions: "Believing is seeing." More attention might be given to identifying cost-effective methods of capturing customer sentiments and designing programs that address (subjective) consumer perceptions.

Figure 1 outlines different Benchmarking Areas: Process Benchmarking, Efficiency and Productivity Indicators, and Customer Satisfaction Indicators. Two other areas with implications for long term performance are often neglected: Financial Sustainability and Water Resource Sustainability. If costs are low because necessary maintenance is deferred, current performance looks "good" but it is purchased at a high price, since future customers will have to bear the costs of current mismanagement. Similarly, water resource sustainability is important for both developed and developing countries: if water sources are becoming contaminated or being depleted, the long term cost consequences are significant. Thus, even the production and cost models described here do not capture all the important dimensions of long term efficiency.

The Figure highlights the variables that are particularly important for obtaining comprehensive productivity and efficiency performance scores for water utilities: inputs and outputs and input prices. The Figure does not capture all the links and feedbacks affecting managerial decisions and performance, but key factors affect sector outcomes. Some variables are under current management's control (like number of employees), while others are the result of past managerial decisions, like the current network (reflecting inherited assets and past maintenance outlays). The cost of capital and the prices of variable inputs (like labor) determine total economic cost associated with different output levels (and service qualities). The roles of governance and organizational incentives are singled out as factors affecting utility efficiency. Due to data difficulties for the cost of capital, analysts sometimes only can identify the determinants of Operating Expenses. Of course, many factors that affect the production process are determined external to the utility (population density, topology of the service territory, customer ability to pay, and access to water resources). Performance scores based on production or cost models need to take such factors into account, so that analysts are comparing apples to apples when evaluating performance. When conducting studies, analysts must address issues related to measurement, data sources, methodologies, and functional forms used to identify key relationships. After reviewing these issues, we discuss problems with company comparisons and describe recent developments in performance evaluation.

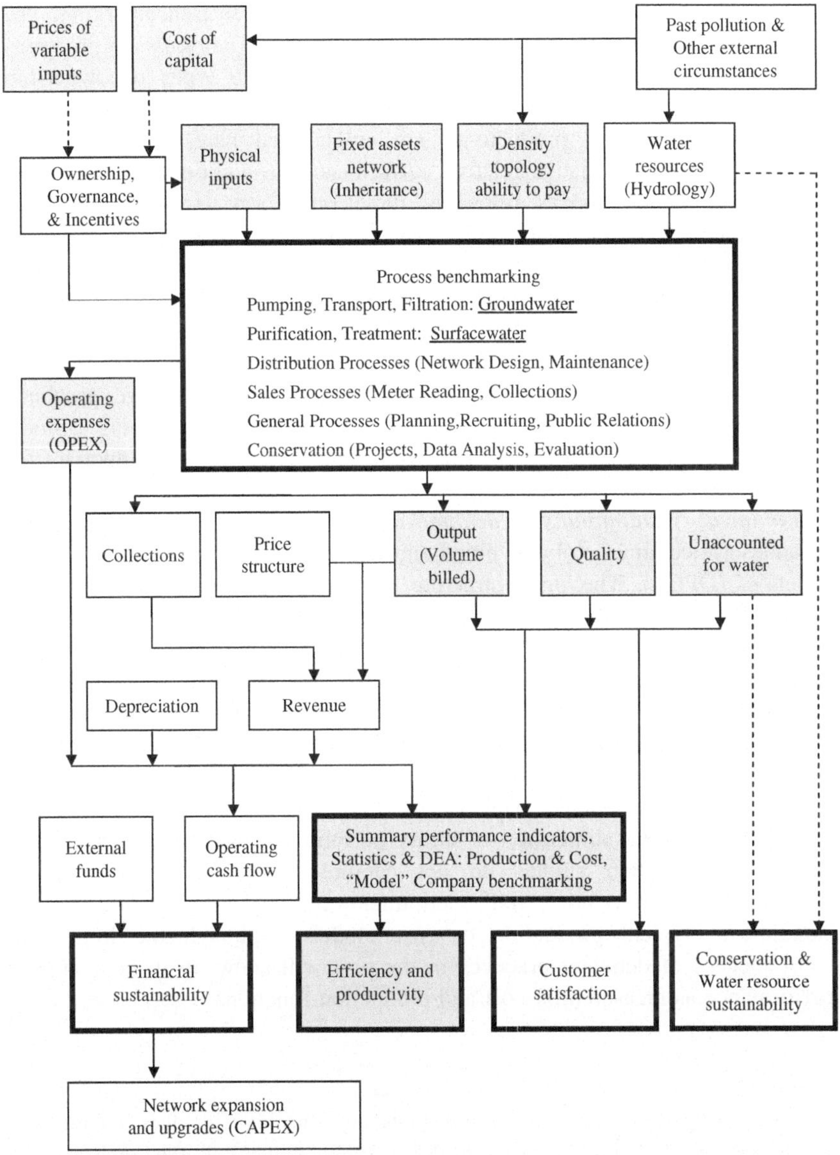

Figure 1. Inputs, Processes, Outcomes, and Performance Benchmarking

Figure 1 includes a box that lists elements of Process Benchmarking. The present volume does not attempt to dig deeply into the various sub-processes that link inputs to outputs. Nor does it examine the types of ownership arrangements and governance systems that affect organizational incentives. Rather, efficiency and productivity are emphasized here, using cost and production models to gauge relative performance. In addition, the bottom of the Figure contains boxes reflecting three other aspects of water sector performance. These dimensions of a water utility's performance are important, but metric benchmarking that uses production functions and cost functions generally will not capture financial sustainability, customer satisfaction, and water resource sustainability. Yet, as has been noted, each of these elements has a significant impact on the long term viability of the water utility. For example, an efficient utility whose revenues fall short of costs will have problems maintaining its network, retaining its best professionals, and attracting capital.[8] Thus, the following highlighted items within the Figure are also important for water utility managers to consider.

Financial Sustainability Benchmarking considers the role of collections, revenues (price times volume billed times percent collected), and operating expenses (OPEX). The price structure includes hook-up (connection) fees, monthly fixed fees, and price per unit consumed (which can involve inclining or declining block prices). Key financial ratios serve as indicators of short term situations and long term performance. Revenues less operating expenses provide the cash flows that facilitate future capacity investments: for both network expansion and upgrades (CAPEX—capital expenditures). Obtaining external funding (either through the issuance of bonds to private investors, to government agencies or to development banks) can be contingent on revenues exceeding OPEX. Clearly, this dimension of utility performance warrants attention, yet cost and production studies do not capture issues raised by financial sustainability. Furthermore, staff recruitment, training, and retention is not captured in typical production or cost models—although the long term consequences of deficient practices in the past will show up in poor current performance as measured by cost and production functions.

[8] Government-owned enterprises still compete for funding with other activities that can benefit citizens. Political leaders will be less willing to invest in water systems if the money is not used in a cost-effective manner. Private investors will avoid public-private partnerships or equity ownership in operations that are poorly managed and/or appear to be financially unsustainable (due to excessive political interference or operating costs exceeding revenues).

Water Resource Sustainability is one issue that universally is given inadequate attention in the analysis of water utility performance. Hydrologists and others modeling water systems can bring deep insights into the implications of water use decisions, including impacts on the environment. Discharges and pollution transportation affect the value of water downstream. In some cases, irreversibilities arise, so it becomes prohibitively costly to restore ecosystems. Nations have developed a variety of institutions to address water sustainability issues. A recent World Bank study decomposed water institutions into three components: water law, water policy, and water administration.[9] The study concluded that the linkages among these institutional features and water sector performance are complex, but that legal factors, enforcement mechanisms, and a sequential strategy for institutional reform can contribute to improved sector performance. Collaboration between agencies responsible for environmental protection, water allocation, and utility service delivery is required for the long term sustainability of water systems. These dimensions of water sector performance are basically public policy issues related to resource management, and are not addressed here.

1.3 Measurement and Data Sources

To identify relative performance in terms of efficiency and productivity, the analyst must draw inferences from observations about inputs, input prices, outputs, and costs. Most economic research is guided by the following:

> *Measurement without theory is wasteful and theory without measurement is generally uninformative.*

Thus, both conceptual frameworks (derived from theories) and quantitative analyses are important if our "map" of reality is to be useful for decision-makers. Some benchmarking studies lack a theoretical basis. For example, an equally-weighted average of six performance indicators does yield a numerical "score" but the implications of the calculated overall indicator is unclear—whether comparisons are made across a number of firms or over time for a particular operator. On the other hand, the application of production theory can be problematic as well: models can become very abstract—characterizing processes in ways that managers might find unclear (or excessively simplistic). This study attempts to bridge theory and measurement; this overview is not

[9] R. Maria Saleth and Ariel Dinar (1999). *Evaluating Water Institutions and Water Sector Performance,* World Bank Technical Paper No. 447, xi–93.

designed for theorists or econometricians.[10] Rather the material is relevant for those translating principles into practice. In addition, process engineers will find the functional forms relating inputs to outputs to be somewhat crude. They will note that coefficients do not direct attention to specific sub-processes that warrant re-engineering. Similarly, financial sustainability, staff development, and customer satisfaction are seldom incorporated into production or cost studies. Nevertheless, this author argues that the methodologies described here can provide important information to those attempting to evaluate the relative performance of water utilities. The data requirements are significant, but generally, the information is a by-product of sound managerial accounting information systems and operating statistics.

Data are necessary to address the following questions:

- What are the key inputs affecting output and cost?
- How do input prices translate into costs?
- What are the basic outputs?
- What are the best measures of water quality?
- How are inputs related to outputs?
- Can specific ratios be used to rank utilities?

Depending on the issue under investigation (and data availability), the analyst will use production or cost functions to identify industry patterns across utilities and/or over time.

The procedures for conducting benchmarking studies must begin with data sources that document company operations. The issues to be addressed include (but are not limited to):

- **Data requirements:**
  - Overall data requirements (which will depend on questions to be answered);
  - Data screening procedures;
  - Information required for analysis (what data need to be collected);
  - Availability of secondary data and where it can be collected.

---

[10] For the most comprehensive introduction to the technical issues associated with benchmarking, see Coelli, Timothy J., D. S. Prasada Rao, Christopher J. O'Donnell, and George E. Battese (2005). *An Introduction to Efficiency and Productivity Analysis,* Second Edition, Springer, xvii–349. A shorter survey is available in a volume by Estache, Trujillo, Perelman, and Coelli (2003). *A Primer on Efficiency Measurement for Utilities and Transport Regulators,* World Bank Institute.

- **Needs for different groups**
  - Utility managers;
  - Water associations;
  - Regulators;
  - Policy makers.

- **Standardization of data to ensure comparability**
  - Definitions;
  - Data verification procedures;
  - Characterization of confidence intervals;
  - Data auditing systems.

- **Methodologies for analyzing data**
  - Data aggregation;
  - Model specification;
  - Time series, cross section, and panel approaches.

- **Institutional capacity and responsibilities for implementing a benchmarking system**
  - Human resource requirements;
  - Costs of establishing and running a benchmarking system;
  - Annual updating of the data;
  - Preparation of benchmarking studies and reports.

Data analysis requires great care, as is illustrated by the following dictum regarding quantitative analysis:

There are three kinds of lies: lies, damned lies, and statistics. (Mark Twain)

Of course, Mark Twain was a humorist and social critic. His skepticism should not deter those who would attempt to conduct sound benchmarking analyses. However, it does remind us that statistical procedures are subject to potential abuse.

1.4 Operational and Accounting Data

Metric benchmarking relies on data from company financial reports, operations reports, and from reports on water quality, where the latter are submitted to (and audited by) public health agencies or environmental agencies. A well-managed firm needs such information for decision-making, although utilities in many countries have not done a good job of collecting consistent data series. Missing data (which may be non-random, since the provider may not wish to reveal particular outcomes), outliers (which may be due to incorrect data entry), measurement errors (stemming from inaccurate instruments or weak accounting procedures), and data validity (the degree to which a measure does, indeed, reflect what it purports to measure) all raise problems for benchmarking.

For simplicity, consider the following sources of information and their applications to two types of models. The framework is very stylized, but illustrates how data from company reports can be used to test models that explain two important operating outcomes:

- Output as a function of inputs, and
- Cost as a function of output and input prices.

Accounting information provides a starting point for many studies, where bookkeeping methods help managers maintain a financial record of business transactions. Managers must prepare statements concerning the assets, liabilities, and operating results of a business. The three main accounting statements are the *Balance Sheet*, *Income Statement*, and *Statement of Cash Flows*. Such financial reports will not provide adequate measures of all inputs, but (over time) they can provide information on trends in network capacity, for example. Such information can be related to *Operational Data* (such as length of the distribution network) to determine whether variables are, indeed, measuring the right things. A utility uses its assets to produce output. For now, the output will be volume of water billed. This output measure can be expanded to include specific indicators of water quality (operational data derived from records). Table 1 illustrates the range of data.

Table 1. Operational Data

Volume produced (water billed, Quantity of sewerage services provided)
Number of connections
Unaccounted for water (Water Delivered to the system less water billed)
Length of Network
Quality of Service (e.g. continuity, proportion of water treated)
Conditioning or Other Environmental Factors (customer density, customer mix, or water source)

Operational data are not necessarily reported in traditional accounting statements, but good managers collect operating statistics and develop asset inventories (or registries) to understand trends and to analyze the effectiveness of engineering activities. The lack of such information is evidence of weak operating controls. The absence of data can be due to low cash flows (stemming from politically driven pricing or weak collections procedures) that make data collection impossible. Or, data gaps may reflect poor management (involving weak governance systems and the absence of internal performance incentives).

Similarly, good managers have regular access to standard financial statements. The balance sheet is a financial statement prepared annually for shareholders (private stockholders or the public—where the Treasury may hold ownership). The Balance Sheet (Table 2) states a company's assets and liabilities. Like a financial snapshot of the company's financial situation at that moment in time, the balance sheet shows the value of the assets as being equal to liabilities plus the net worth of the company. Assets (such as buildings and pipes) are funded from cash flows from previous years (cumulative retained earnings) or from the issuance of debt (and equity, in the case of private utilities). If funds are "gifts" (grants) or involve no liabilities, then they become part of the net worth of the entity. A debt used to fund projects (as in issuing a bond) encurs future obligations: a stream of interest payments and the principal is expected to be repaid. Such debt represents a long term liability.

Table 2. Balance Sheet: Assets, Liabilities, and Net Worth

Assets	Claims on Assets
Current Assets	Current Liabilities
Cash	
Accounts Receivable	
Long Term Assets	Long Term Liabilities
	Owners' Equity (Retained Earnings)

Note that Accounts Receivable is viewed as a current asset—so long as there is a reasonable expectation of payment. However, uncollected billings are extremely high for water utilities in some countries. If the economic reality is that the bills will never be paid, they should be "written off" (thus reducing owners' equity). Even state-owned enterprises should have some measure of owners' equity, since taxpayer/citizens are the ultimate owners.

We can think of using the long-term assets (pipes and pumps) operating in conjunction with variable inputs (like labor or electricity) to produce a specific volume of output. Some studies include bulk water delivered to the utility as an input (to capture water losses). For now, consider a very simple production function:

Production Function: Output, **Q** (Quantity or Volume Billed), is a function **(F)** of Inputs

Q = **F** (Input Quantities) = **F** (network assets, labor)

The production function relates physical quantities of inputs to physical quantity of the output (here, quantity—volume of water delivered to customers).

A similar relationship can be established between costs and output level and input prices:

Cost Function: Cost, **C**, is a function (**G**) of Output and Input Prices

C = **G** (Output, Input Prices) = G (Q, wage rate, energy price, cost of capital)

Total cost would include some annualized cost for the network capacity, which might be calculated as depreciation plus interest payments or by the cost of capital times the amount of capital used in production. A component of total cost, like OPEX, would be affected by the quantity and quality of the output and input prices. Sometimes analysts study components of costs (like pumping expenses or energy costs). Such studies try to identify key cost drivers and to determine whether particular utilities have excessively high costs. A detailed version of the Simplified Income Statement (Table 3) contains some of the information required for estimating cost relationships.

Table 3. Simplified Income Statement

Net Sales (total revenue = price times quantity times percent collected) − Operating Expenses (e.g., labor expenses) − Depreciation
= Operating Income
− Interest Payments
− Taxes
= Net Income

Thus, an equation that shows how the determinants of cost relate to total cost can provide insights regarding how well a utility is performing relative to comparable utilities. If two utilities produce the same amount of output but have very different costs, the differential performance may be due to a number of explanatory factors, including inefficiency.

Another Financial Statement that sheds light on the sustainability of the utility is the Statement of Cash Flows (Table 4). It reconciles the initial and end-of-year cash on the firm's Balance Sheet.

Well managed utilities have substantial data that can be used to monitor performance. Although this volume does not emphasize the Statement of Cash Flows, this financial statement provides essential information for evaluating the financial sustainability of the utility—a potential area for benchmarking.

Table 4. Statement of Cash Flows (Simplified)

Cash at Beginning of the Period
+ Net Operating Profit Less Adjusted Taxes [= EBIT (1−tax rate)]
+ Depreciation (a non-cash expense)
= Cash Flow From Operations
Cash Flow From Operations
− Investment Outlays (CAPEX)
− Interest Payments
− Dividends or Transfers to the Treasury
+ (−)Funds from Issuing (or redeeming) Bonds
= Cash at the End of the Period

1.5 Illustrative Functions: Model Specification

To illustrate specific functional forms for production and cost functions, some stylized relationships are presented below. Clearly, obtaining good measures of relative performance is contingent on using methodologies that yield quantitative relationships that reflect reality. Let us begin by considering a function (recipe) that shows how the level of output depends on the level of an input.

1.5.1 Production functions

The production function, F, can be viewed as a recipe (or formula) that translates some level of inputs (L) into an output level or quantity (Q). The particular coefficients of the model (or parameters of the model) specify how the inputs relate to the output. Assume that the network is built, that bulk water is costless, and that output (billed water) only depends on labor. Thus, we might believe that the following relationship captures reality over the relevant output range:

$$Q = 10\,L - .2\,L^2,$$

We have the following potential outcomes: as L increases, volume increases:

Q	L
0	0
9.8	1
19.2	2
28.2	3
36.8	4
45.0	5

As L increases, output (Q) increases according to the production function formula. The parameters of the production model determine how the volume of water billed (the output) depends on the amount of labor, L. The formula indicates how L affects Q, where the coefficients, 10, .2, and the 2 (for L squared) would be estimated from statistical or engineering data. Of course, estimating the parameters can be a difficult process requiring the use of sophisticated quantitative techniques.

Furthermore, if the function is *actually*

$$Q = 12L - .3L^2,$$

the output numbers in the Table above would be "wrong." If L = 4, output would be 43.2 NOT 36.8 as stated in the Table. So correctly estimating the parameters of the function is essential if the model is to accurately reflect reality. Otherwise, a firm might be identified as being relatively efficient when it is not, or as inefficient when it is, in fact, on the frontier. Here, the operational reality under investigation is how the output level depends on the level of inputs. If we knew that the original function was "correct" and a firm was actually producing 20 units of output with 5 units of labor, the analyst would conclude that the firm was not operating efficiently (assuming that the model has captured all the relevant factors affecting production).

Multiple outputs (water and sewerage services), service quality, multiple inputs, and other factors can be incorporated into the empirical analysis through a variety of techniques. For now, we retain this simplistic framework to illustrate the key concepts requiring quantification. Issues of model specification and interpretation will be discussed more thoroughly later in this volume.

1.5.2 Cost functions

Water utilities purchase many inputs to produce water, and water quality can differ across utilities. In some cases, data on the network itself (including capital inputs and the cost of capital) are unavailable or flawed, leading to studies that focus on the relationship between operating expenses and output. Thus, we will consider a variety of cost functions later. In addition, the variables used to approximate input prices and output will be discussed later. In some cases, proxies can be used to approximate the appropriate values. For now, we are including neither unpaid bills, nor many other factors (such as unaccounted for water). The production of water/sewerage service (WSS) can be incorporated into multi-product cost models, but discussion of such extensions is deferred until later. A simple model would relate cost to input prices and the level of output.

Cost Function: Cost (C) is a function (G) of Input Prices and Output (Q)

$$C = G\ (\text{input prices, output}) = G(w, r, Q)$$

where w = price of labor, and r = price of capital

The benefit of estimating a cost function is that managers often think in terms of cost and revenues. Interpreting the results in an intuitive manner is valuable. Furthermore, sometimes categories of cost are related to output characteristics and to conditioning factors (like population density). For example, the water regulator for England and Wales (OFWAT) has used statistical models to determine whether specific types of expenditures by water utilities were excessive for achieving particular output levels.

It should suffice to note that the production and cost frameworks have different data requirements and can provide information on different operating features of particular water utilities. The relationships will be depicted in Figures in subsequent sections.

Factors beyond Management's Control: Analysts have used several methodologies to estimate coefficients of production-functions and cost-functions. The production function framework can help identify utilities producing less output than comparable utilities using the same amounts of inputs. Or, in the case of cost functions, the framework allows the analyst to find utilities with higher costs than predicted by the model (for the output level produced by the water utility). Interpreting production or cost models requires great care because alternative functional forms (model specifications) and different variables can yield different results. The resulting efficiency scores can be sensitive to such factors. For example, cost (and resource requirements) depends on the source of the water (surface or subsurface), its distance from the consumers, the population density, and other conditioning factors. Such factors are generally beyond management's control, so they should be taken into account when evaluating relative performance. The strengths and limitations of alternative methodologies will be outlined shortly. However, one point should be clear:

The conclusions one draws from quantitative studies are only as good as the underlying data and the model specifications.

Studies have addressed the following types of questions:

- What have been the performance trends for utilities over time? Such studies generally utilize a set of Core Indicators, often aggregating them into some Overall Performance Index (OPI).

- To what extent are increasing returns to scale present in supplying water? Such studies tend to utilize more sophisticated statistical analysis. These studies can have implications for creating incentives that promote regional consolidation (or coordination) if the scale economies are shown to be substantial.
- To what extent are there economies of scope in the provision of water and wastewater services? Multiproduct cost functions are more complicated to estimate, but if analysts are trying to determine whether single product or multiproduct firms would be most cost-effective, empirical studies can help identify patterns.
- Are privately owned utilities more or less efficient than publically-owned utilities? Here, a cost function could include a shift variable to determine whether costs tend to be higher or lower for particular ownership arrangements (which tend to have different performance incentives).
- Does regulation or benchmarking affect utility cost performance? Again, models can be specified that attempt to identify the impact of regulation, holding all other factors constant.

In order to gain appreciation for the level of specification utilized in practice consider example a Portuguese cost benchmarking study by Correia, Brochado, and Marques reported in Cabrera and Pardo (2008).[11] The cost model utilizes the following framework:

$$TC = C\ (W, L, S, K, E, CU, N, LO, dP, dO, dC)$$

The variables are defined as follows:

TC total cost of water production is a function of,
W distributed water volume,
L wage rate (unit staff cost per employee),
S cost of imported water (obtained as the ratio of cost of goods sold and the volume of distributed water,
K capital cost (calculated by cumulative capital expenditure CAPEX divided by network length),
E energy cost (estimated as the ratio of current costs and the volume of distributed water),
CU number of customers,

[11] Correia, A. Brochado, and R. C. Marques (2008). "Benchmarking the Performance of Portuguese Water Utilities," in Cabrera and Pardo (2008), pp. 273–283.

N network size (measured in terms of the lengths of mains),
LO water losses value (measured two different ways, in cubic meters and as a percent),
dP dummy variable for the type of management,
dO dummy variable indicating provision of other services besides water supply, and
dC dummy variable for the existence of water abstraction facilities.

In statistical terms, total cost of water production (TC) is the dependent variable, and the other variables affect that cost: they are independent variables. The Portuguese Team transformed the values into natural logarithms[12] in order to estimate the cost function and analyze the impacts of changes in the input variables on cost (the dependent variable).[13] They used a fairly sophisticated form of regression analysis called Stochastic Frontier Analysis (briefly summarized later). For now, it is enough to note that the model focuses on those firms that have the lowest cost for given values of the explanatory variables. So efficiency scores are based on performance relative to firms on the cost frontier. For their sample of sixty-six water utilities (serving 60% of Portugal's population in 2005), efficiency scores ranged from .45 to 1, with the mean and median scores being about .85 for Model 1. Utilities on the cost frontier have a score of 1. Later, this volume explores the implications of such rankings for managers and regulators.

In this particular study, all the explanatory variables were significant and had the expected signs, except water losses. The researchers explain the result as indicating that reducing water losses requires greater maintenance and rehabilitation expenses, so higher losses involve lower current costs. The study finds that as the wage rate increases one percent, total costs increase .08 percent—indicating a small role for labor costs.

---

[12] The functional form is called a Cobb-Douglas function (in honor of the two economists who first utilized it). Without going into technical detail, the resulting equation is estimated in log-linear form, so the coefficients can be interpreted as elasticities. The coefficient shows the percentage change in cost from a one percent change in the value of the independent variable.

[13] They estimated two models, Model 1—where water losses is measured in cubic meters, the same dimension as W, the total water volume distributed (the output) and Model 2, where the water losses variable was measured as a percent of volume distributed. An alternative specification could have made "water delivered" the output variable (instead of water entering the distribution system) and included volume of water losses as an "input," to check whether water losses increased or reduced costs.

As the volume of distributed water increases 1 percent, costs increased .51 or .62 percent (depending on the model). This result indicates substantial *short run* output economies, holding network size, number of customers, and other variables constant. This should not be surprising, since so much of the costs are being held fixed. *Long run* economies (or diseconomies) of scale relate to how costs change when the output, number of customers and network size increase together proportionately. The researchers find diseconomies of scale in the water sector, concluding that this reflects "… the fact that the larger companies usually have higher staff costs, do not need specialized manpower or because … features of the services lead to an increase in production that requires a new [distant water] source …" (p. 282).

Other results from this analysis of the Portuguese dataset suggest additional interesting patterns. The values of the dummy (cost shift) variables suggest that private utilities have lower costs, providing other services lowers water costs (indicating potential economies of scope), and operating one's own water abstraction facilities lowers costs (perhaps because this element reduces the cost of imported water). Rather than going into further detail here, it is important to note that more than one-hundred such studies have appeared in the academic literature, in journals like the *Journal of Regulatory Economics*, *Utilities Policy*, and the *Journal of Productivity Analysis*. Some utilize production function models rather than cost functions, but the two types of studies ought to yield comparable rankings of utility performance. Both shed light on how different factors affect costs and productivity within (and across) water utilities. Some of these studies have been used in designing incentive schemes that put pressure on poorly performing utilities.

The study summarized above is meant to help water managers recognize the information technology and data analysis skills necessary for a comprehensive benchmark initiative. Of course, professional consultants and/or academic researchers are likely to be in the best position to determine input variables and the functional form in an empirical study. Nevertheless, government agencies and utility associations are certainly able to develop such skills internally. Furthermore, only a few production and cost function studies by academics focus on relative performance of firms; many studies have other purposes, like determining the extent of scale economies in the industry.

1.6 Company Comparisons

Company comparisons allow analysts to gauge relative performance across firms and evaluate absolute levels of performance over time. Thus, performance indicators are used to rank companies in terms of efficiency. Such indices or scores can then be used by regulators for establishing targets. In addition, governance procedures within companies can incorporate this information into managerial incentive packages. Rankings can serve as catalysts for better stewardship of water and other resources. However, care must be taken to use comprehensive indicators, lest those being evaluated "game" the system, improving performance on a subset of indicators, and letting other dimensions of performance fall.

Partial comparisons can provide a useful starting point for identifying poor performers. *Specific Core Indicators* such as accounts receivable (uncollected billings) or volume billed per worker can be utilized in the early stages of infrastructure reform. If a water utility is under severe cash flow problems, quality of service tends to suffer and it will be difficult to obtain funds for network expansion. Thus, partial indicators can be used to set targets—providing rewards to managers (and utilities) who achieve those targets.[14] Simple trend analysis can provide a first step in determining the extent of the problem.

Relative rankings allow the different groups to compare the performance of utilities in comparable situations. Here, the key problem is how to select firms that are truly similar to one another. Alternatively, how can the rankings correct for (or incorporate) the different conditions managers face? Analysts want the relative ranking to reflect managerial decisions rather than the unique characteristics of service territories beyond managers' control, including topography, hydrology, and customer density. In addition, history matters: current managers have inherited utility systems that reflect a set of political and economic (including regulatory) decisions made by others. Thus, performance improvements over time also need to be taken into consideration.

Absolute comparisons are also necessary, since the weakest performer in one group might have much better performance than the best firms in another group of comparable firms (say, those in another country at a similar stage of development). Comparisons are valid so long as the results do indeed tell us whether particular firms are performing below their potential. Therefore, the problem of consistency requires a reasonable, reliable, and stable scorecard based on the political and economic environment of different countries, the specific targets set by the regulators, the availability of the data, and other environmental factors. In addition, trends over time can tell observers whether managers are improving performance.

---

[14] Mugisha, Silver, Sanford V. Berg and William T. Muhairwe (2007), "Using Internal Incentive Contracts to Improve Water Utility Performance: The Case of Uganda's NWSC," *Water Policy,* 9(3): 271–284.

Comprehensive comparisons are required if indicators of relative performance are to be meaningful. For example, comparisons should not ignore water quality, such as continuity (hours of service per day) or microbiological and chemical quality. In a benchmarking cost study that omitted quality, "low-cost, low quality" companies may be labeled as "efficient" companies: such a conclusion would be quite inappropriate.[15] Earlier points about water resource sustainability and financial sustainability underscore the importance of making comprehensive comparisons across utilities.

Now that benchmarking terminology has been outlined, we can turn to a Benchmarking Checklist that will help regulatory analysts and utility managers more effectively conduct and evaluate benchmarking studies. The checklist should serve as a template to ensure that key issues are not swept under the rug.

[15] In addition, quality can be an important issue in Total Factor Productivity (TFP). For instance, Saal and Parker (2001) show that the TFP change in the U.K. water sector has been extremely slow in recent years, but the quality has improved significantly because of the large increases in minimum standards, which required significant outlays. Thus, the use of unadjusted TFP change measures during this period would understate the actual TFP improvements. D. Saal and D. Parker (2001). Productivity and price performance in the privatized water and sewerage companies of England and Wales. *Journal of Regulatory Economics*, 20(1): 61–90.

Chapter 2 Checklist for Conducting Benchmarking Studies

In considering best practice, this book draws upon the experiences of participants in a 2007 PURC/IADB Benchmarking Workshop (San Jose, Costa Rica).[1] Regulators and managers identified factors having an impact on data quality, data collection, benchmarking methodologies, and possible policy and regulatory implications of performance rankings:

(1) The person (and the company) responsible for the data (collection, verification, storage, and processing) must be convinced about the importance of his (or her) role. Besides serving as a report to the regulator or to any external institution, data must be viewed as important and useful for the company – for strategic, operational, administrative and commercial purposes.
(2) Duplication of data storage files inside the firm and data reports (in specialized formats) to external institutions must be avoided or reduced. Duplication raises administrative costs and opens up the possibility for little "Information Empires" where individuals exercise power by withholding data from those who should have access to information.
(3) There needs to be a person responsible for data within the company. Professionals committed to benchmarking programs bring expertise and energy to this difficult task. However, the formal position (in addition to the person) must have continuity over time. This formalized role is needed to address internal turnover problems which limit data

[1] These points are developed more fully in "Water utility benchmarking for managerial and policy decisions: Lessons from developing countries," with Maria Luisa Corton, in *Performance Assessment of Urban Infrastructure Services*, eds. Enrique A. Cabrera, Jr. and Miguel Angel Pardo, London: IWA Publishing, 2008, 307–320.

collection; lack of continuity and commitment causes gaps in time series and in cross-section observations.

(4) Clear variable definitions allow outsiders to interpret information; consistency and clarity are fundamental to the management process. IBNET and IWA initiatives have greatly improved our ability to develop comparable data across entities.

(5) Factors external to the company may have an impact on the collection and storage of data. For example, the existence of records and maps of the city; the frequency of the country census; municipal or city restrictions regarding the network design; number of connections per km, and type of users of the network.

(6) Data disaggregation improves decision-making. Clear customer classification (residential, industrial, and commercial) allows for more accurate information regarding operational performance of the company. In addition, maintaining data series on particular regions or divisions of a company allows top managers to develop strategies for rewarding strong performance. Disaggregated data allows managers to target areas of sub-standard performance and facilitates quantitative studies of cost and productivity.

(7) Better operational data collection procedures are needed: timely reports that identify patterns mean that network repairs can be addressed in a comprehensive and cost-effective manner.

(8) Information technology is necessary, but not sufficient, for sound management: information systems should link financial-commercial-operational data. Leaders can only manage what they measure.

(9) The company information needs to be public to promote managerial accountability and citizen confidence in infrastructure services. Even rough comparisons can put pressure on political leaders to fulfill promises to provide funds for network expansion and on managers to deliver services at least-cost.

(10) Overall, there is a need within the firm for more information regarding benchmarking methodologies and their application. Larger water utilities have engineers who are familiar with process benchmarking. However, there is a need for capacity building in the area of metric benchmarking—starting with trends in Core Indicators, and moving to basic statistical reports and DEA studies.

Building on these lessons, and on the experiences of practitioners in other nations, the benchmarking process can be divided into five steps (as shown in the flow chart, Figure 1) the: identify objectives, select methodology, and gather data; screen and analyze data; utilize specific analytic techniques; conduct consistency/sensitivity tests; and develop policy implications. Each step includes a set of sub-steps, described in the remainder of this section.

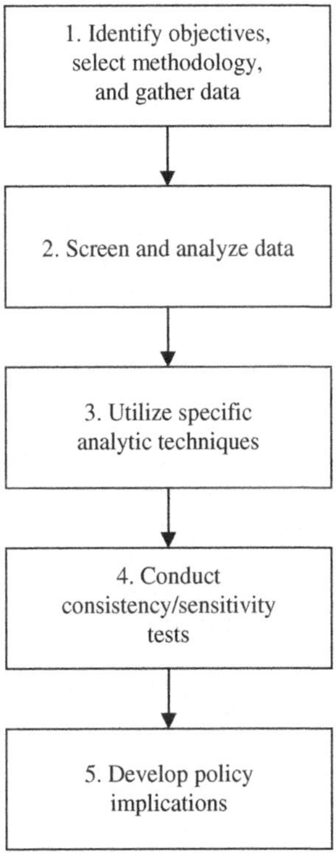

Figure 1. Five Steps of Benchmarking

2.1 Identify Objectives, Select Methodology and Gather Data (Step 1)

Analysts must make choices regarding the issues to be addressed, the time period to be studied, and the types of comparisons to be made. These choices will reflect current analytic capabilities, an initial understanding of data availability, and preliminary methodological choices. The objectives of any benchmarking study will depend on the most important policy issues under consideration. Figure 2 below outlines the initial stages for initiating a benchmarking study. Performance benchmarking is highlighted, but other broad methodologies are also noted for completeness.

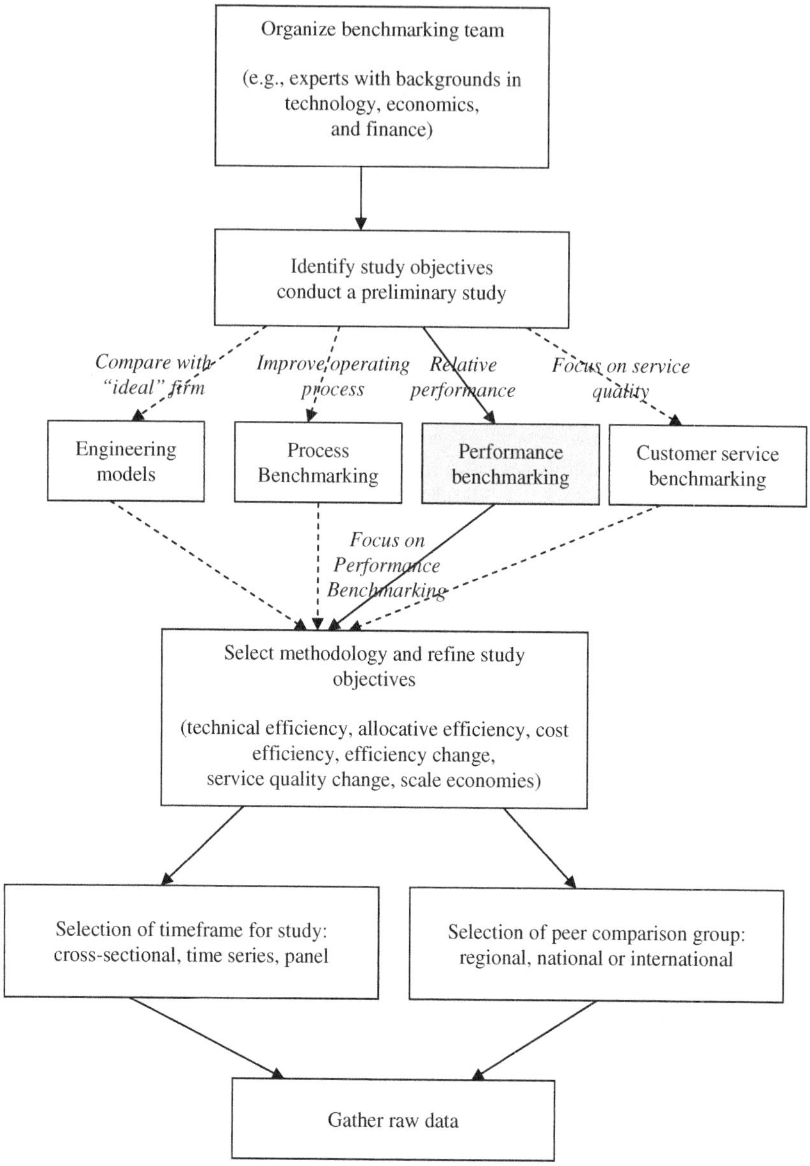

Figure 2. Step 1: Identify Objectives, Select Methodology, and Gather Data

2.1.1 Organize benchmarking team

It is important to involve staff professionals who have a deep understanding of the operating characteristics and technology of the industry being benchmarked so the team can select the most suitable variables and indicators.

- Are there staff members with finance and accounting backgrounds to help construct financial statements and check the financial status of firms?
- Does the team include members with training in econometrics and statistics who have knowledge of conducting surveys, testing empirical models, and analyzing production processes?
- Are there professionals who understand the regulatory and organizational issues who can help interpret the results and develop policy implications?

Areas requiring specialize skills include the ability to characterize regulatory mechanisms, governance (under public, private, and mixed ownership), and managerial incentive contracts. If these differ across operators, statistical research may be able to determine the impacts of different institutional arrangements. At a minimum, such external factors must be controlled for (as in the Portuguese study summarized earlier).

2.1.2 Identify study objectives

The first task for the benchmarking team involves identifying the study objectives and deciding whether processes or performance will be benchmarked. Initial benchmarking initiatives should be simple, only require basic data, and be defensible. The approach should be based on a thorough review of the problems and opportunities within the industry, based on preliminary discussions with stakeholders, including the organizations under review. For water utilities, issues might include productive inefficiency, poor service quality, severe water loss, poor operation management, financial distress, and low coverage ratio (population receiving service divided by total population or percentage of households with access to water).

- Which problems are most urgent and most important?
- Is there enough information about each of these problems, or is additional study and research needed?

The work plan should identify items that are most critical to a successful analysis. Based on the status of the industry and the team's strategic plan, decision-makers can make an initial choice from among the four broad types of benchmarking studies identified so far: Engineering/Model Company, Process Benchmarking, Performance (Metric) Benchmarking, and Customer Survey Benchmarking. Other areas noted earlier (financial and water resource sustainability) could also be investigated.

2.1.3 Select methodology and refine study objectives

After making the initial choice (here, assumed to be Metric Performance Benchmarking), decision-makers need to focus and refine this choice to identify specific aspects of the practice they want to benchmark.

- To what extent are data on inputs, outputs, and environmental conditions available on a consistent basis across organizations?
- How do the performance comparisons relate to the objectives of those conducting the study?

To answer these questions, analysts need to have a clear idea of what kind of information they are seeking, what aspects of performance are likely to need improvement, and what targets for future performance might be most reasonable. For instance, if the analyst chooses to conduct a performance benchmarking comparison, he or she will need to decide the focus of the study:

- Are firms producing on the frontier—so no additional output could be obtained from the inputs? (technical or engineering efficiency)
- Is the right combination of inputs chosen to minimize costs? (cost or production efficiency)
- Is the quality of service consistent with customers' willingness to pay? (an element of allocative efficiency)
- Do prices reflect incremental production costs? (an element of allocative efficiency)
- Are organizations operating where scale economies have been achieved? (scale efficiency)
- Are organizations operating where economies of scope have been achieved? (scope efficiency reflecting multi-product economies, such as delivering both water and wastewater services)
- Are there multiple issues to be addressed? (If so, such complexity is likely to require a combination of approaches)

Furthermore, analysts might be interested in the total factor productivity change (TFPC), which can be further decomposed into technical efficiency change, scale efficiency change, technology change, quality change, and allocative efficiency change. Analyzing such decompositions involve substantial data requirements: this activity might be a low priority for a developing country, but a higher priority issue for a developed nation. At this stage, analysts need to have clear objectives so appropriate data are collected and a suitable model can be specified.

2.1.4 Selection of timeframe and peer comparison group

Based on the objectives, decision-makers then need to choose the timeframe and comparison group for the analysis. For instance, if the objective is to compare the relative efficiency of the water utilities in a particular year and the sample size is reasonable, analysts can conduct cross sectional analysis. If the objective is to evaluate the efficiency change of a single utility over time, analysts can use annual data from the utility for time series analysis. In this case, a partial indicator might be more suitable than frontier methods because of the limited sample size. If the objective is to evaluate both the current period relative performance and the TFPC, a panel data analysis is necessary, consisting of repeated observations on the same cross sections of firms over time.

Regarding the selection of peer comparison groups, a number of options can be considered:

- **Within company comparisons:** Most utilities are divided into divisions serving different geographic areas. Data from each division can be used to identify strongest and weakest performing divisions. Managers can use the results in developing targets and compensation packages. Regulators can use the results to gauge how the entire organization might improve performance.
- **Within-country comparisons**: A national utility regulatory task force may want to compare the relative performance of the utilities within the nation or a particular region (where density, topology and hydrology may be similar).
- **Cross-region comparisons**: On the other hand, analysts and policy-makers may want to compare the utilities within a region to the utilities in other regions to check for efficiency gaps.
- **National comparisons**: Similarly, a national regulatory committee may want to conduct national or international benchmarking to evaluate the relative efficiency of domestic utilities or efficiency gaps between domestic and foreign utilities.

However, there are also potential obstacles and risks to conducting cross regional/international benchmarking.

- **Data availability:** For a regional regulator, it might be difficult to get detailed information on utilities in other regions because the regulator/ utilities in other regions may be reluctant to provide information.
- **Data consistency:** Similarly, it might be difficult for the national regulatory committee to obtain consistent data from other countries (due to data definition problems).
- **Exogenous factors:** A more important issue is the potential difference among the countries, such as geographical environment, population, weather, and past natural disasters.
- **Accounting definitions:** Furthermore, exchange rate adjustments (for cost comparisons) and different accounting standards complicate international benchmarking.

If a regional/national regulator is not aware of these potential differences, the benchmarking results might be biased or incorrectly interpreted. For example, in the area of financial accounting, different countries may have different definitions of revenue, net income, operating cost, capital expenditure, and depreciation, which might exert significant effects on the benchmarking results. Therefore, for international benchmarking, it is safer to use the physical inputs/ outputs instead of monetary inputs/outputs. IBNET, IWA, and regional collaborative groups have a number of initiatives to improve data collection and reporting procedures.

Overall, benchmarking at the local, national, and international levels helps all water and sanitation utilities identify and share best practice and new knowledge. Collaboration can promote the delivery of the best services for customers. Whatever their development status, groups can benefit from continuing professional education and cooperation. Regional and international initiatives facilitate the creation of international data banks that provide a productive environment for international benchmarking studies. However, it is important to pay attention to the potential differences among countries when conducting international benchmarking studies.

2.1.5 Gather raw data: collection issues

After deciding the objectives and scope of study, the more comprehensive data collection can begin. Here we consider the main problems experienced by water utility companies in developing countries, which have generally experienced poor performance and low productivity growth. Risks associated with each problem will be identified to illustrate the tasks facing regulators, government ministries, operators, multilateral organizations, and private investors.[2]

Technical and operational problems

- Inefficient operational practices (often reflecting poor governance systems),
- Inadequate maintenance (stemming from under-pricing service in many jurisdictions),
- High unaccounted-for water (caused by leakage and commercial losses—theft),
- Limited service expansion (based on financial constraints), and
- Inadequate water quality procedures (both for monitoring and producing potable water).

How does an outside analyst evaluate technical and operational risks associated with these problems? Unless a government-owned utility keeps adequate records there will be insufficient information about the state of installations and the need for replacement, rehabilitation, or expansion. Furthermore, operational performance of the system is questionable. How old are pipes and pumping devices? Have maintenance procedures been adequate in the past? What is the percentage of water not accounted for? What is the source of losses—leakage or commercial losses (theft)? What procedures are used to guarantee water quality? What is the level of operating costs? What are system expansion possibilities? Are water sources adequate for the long term? This litany of data questions illustrates the types of uncertainties facing those who analyze how inputs translate into outputs or production costs at different stages of production.

Commercial and financial concerns

- Un-metered systems create distortions in consumer charges,
- The amount of water produced is often *estimated* (instead of being based on actual measurements),

---

[2] Parts of this section draw upon Berg and Corton (2002), "Infrastructure management: applications to Latin America," in *Private Initiatives in Infrastructure: Priorities, Incentives, and Performance*, eds. S.V. Berg, M.G. Pollitt, and M. Tsuji, Edward Elgar: Cheltenham, UK.

- Reliable consumption (or billing) data may be unavailable because of poor customer recordkeeping,
- Inefficient billing and collection practices lead to high levels of accounts receivable, which translates into non-payments,
- Laws may prohibit cutting off water services, dramatically reducing customer incentives to pay bills,
- Government agencies may not pay their bills, placing a burden on other customers to cover costs,
- Realized revenues may not be sufficient to generate funds needed to expand service and protect water resources and the environment from contamination,
- Tariff policies often do not reflect the true economic cost of future water supplies, and
- Tariff structures with large cross-subsidies characterize many utilities (pricing below cost leads to utility disincentives to serve those customers who are officially targeted as "worthy" of subsidization). Others (generally those who are relatively well off) end up benefiting from the subsidies.

Cost recovery cannot occur when there is such uncertainty regarding these aspects of operations. The baseline is always current procedures and capabilities, so if these are inadequate, organizational changes will be required. New administrative procedures are likely to be disruptive, requiring internal education and buy-in by current employees and customers. For example, what are the mechanisms for responding to customer complaints? Sometimes it is easier to start from scratch than to try to graft new systems onto old procedures. Similarly, historical records are required for making forecasts. How has demand evolved over time? Are the seasonal and hourly consumption patterns predictable?

In the area of financial risks, currency valuation and convertibility raise issues. Mechanisms for hedging risks have a high priority for external investors who need to be aware of government rules regarding remittances by foreign companies (and likely developments in this area). If the utility is publicly-owned, it will still pass risks onto customers, so managers still need to develop strategies for risk-mitigation.

Human capital and personnel issues

- Excess staff indicates poorly managed water utilities,
- Political appointments and intervention are present in many nations,

- There is often an inability to attract managerial talent and qualified technical staff due to lack of adequate incentives (or caps on civil servant salaries), and
- Frequent turnover of high-level staff combined with low productivity (and lack of discipline in the labor force) present problems for managers.

Regulatory monitoring by public agencies and due diligence by potential investors both require investigations into a number of issues, including the way contractual disputes are resolved:

- Is a union contract in place?
- Have pension fund contributions been kept up?
- What is the managerial compensation scheme?
- Are current managers the result of political appointments?
- What is the turnover of staff at various levels of responsibility?
- Are job descriptions flexible?

Most of these issues will not be revealed through statistical studies of relative performance, yet they have implications for the performance of the water utility.

Regulatory governance and incentives issues

Financial markets view the regulatory regime (and political environment) as major determinants of the likely riskiness of cash flows. In addition, government funds have opportunity costs: education, health, transportation and other infrastructure areas compete with water utilities for funds. Private investors and firms with managerial contracts seek a number of features in the environment to insulate decisions from day-to-day politics, while ensuring long-run sustainability of the regime itself.[3]

- What are the prospects for performance enhancement?
- How has regulatory governance (procedures used by the agency to conduct its activities) affected transparency and citizen participation in the implementation of public policy?
- What monitoring systems and internal incentives are utilized by Boards of Directors?

---

[3] King and Pitchford (1998) provide an overview of privatization issues. S. King and R. Pitchford (1998). "Privatization in Australia: understanding the incentives in public and private firms," *Australian Economic Review*, **31**, 313–328.

Beyond the issue of private participation, the role of international organizations and donor nations depends on expectations regarding reform. Multilateral organizations are hesitant to support water systems that have no prospects for performance enhancement (and the lack of data means that baselines cannot be developed). Output Based Aid from international funding organizations and national agencies is predicated on setting and meeting performance targets. Appendix 1 (Variable Definitions) lists some variables that have been used in benchmarking studies to control for private and public ownership, regulatory incentives, and corporate governance.

2.2 Screen and analyze data (Step 2)

Conducting a benchmarking study is an iterative process. Detailed screening of the data will result in greater refinement in the timeframe, sample size, and statistical techniques. The sub-steps of this second stage require more technical skills than the initial framing of the issues to be investigated. However, feedback should be elicited from those who participated in Step 1 so that there is agreement regarding decisions taken at this stage.

After assembling the raw data, the benchmarking team should screen the data carefully to ensure the quality and quantity of information being gathered meet the requirements to allow for a successful project. This process is crucial to the study because poor data quality (inconsistent definitions, missing data or extreme data values) may lead to biased results. Insufficient data could result in the use of more limited models that might constrain functional forms so as to yield biased or skewed results. Figure 3 identifies four broad metric methodologies that might be utilized.

The four broad methodologies include:

- *Partial indictors,* summary indices sometimes aggregated into *core overall performance indicators* (OPIs combine partial indicators of operating or financial performance),
- *Total factor productivity indices* (an index number approach that considers output per unit input—where multiple inputs are taken into consideration to gauge efficiency levels and changes),
- *Relative performance using data envelopment analysis* (a non-parametric technique that makes no assumptions about the functional form of production or cost functions),
- *Relative performance using statistical techniques* (parametric approaches that involve assumptions about functional relationships within production and cost functions).

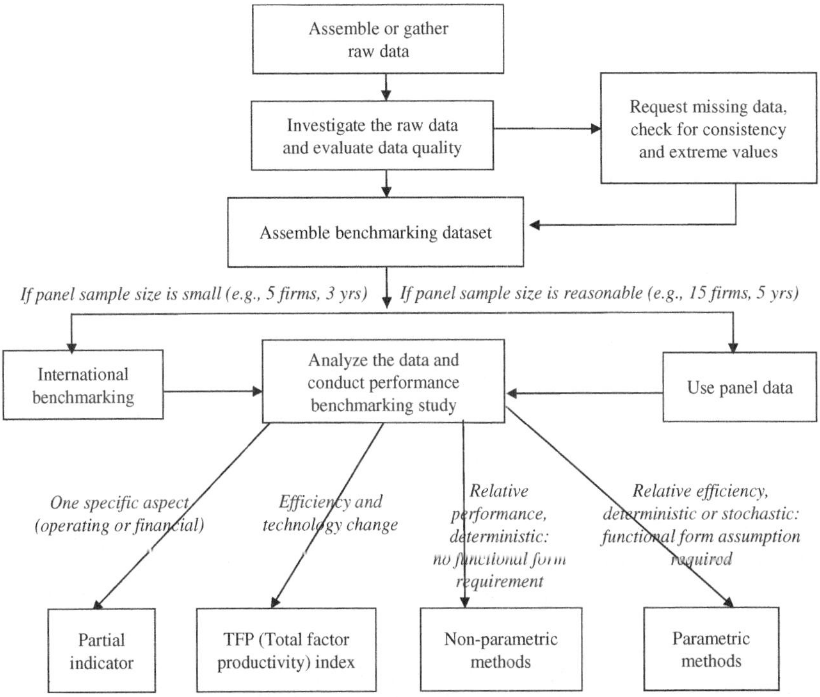

Figure 3. Step 2: Screen and Analyze Data

Each of the above approaches has strengths and limitations, which will be discussed in a subsequent section. Some argue that the limitations of the methodologies supports postponing studies until the data can be fully audited and analysts achieve more agreement regarding which methodologies should be applied in particular circumstances. Others maintain that we can learn a great deal though preliminary analyses, and that the data collection effort will be strengthened by research initiatives. Those who prepared this study are part of the second camp.

2.2.1 Investigate the raw data and evaluate data quality

Measurement is one of the first steps in the benchmarking process. Making comparisons requires significant amounts of data that are often quite difficult to collect. The results depend on the accuracy of the data that are collected. The issues are present from the standpoint of an external utility regulator, but similar issues arise in managerial self-studies. The information collected may be verified by the regulator (or analyst), but this usually comes at some cost. In addition, some data are unverifiable by the regulator or extremely costly to acquire. From this point of view, the regulator has to trust the firm regarding the truthfulness and accuracy of reported data on the operations of the utility: inputs, input prices, outputs (in volume and quality), revenues, and status of assets. The regulator cannot know for sure whether the reported operating expenditures and capital expenditures reflect the least cost for achieving reported outputs. We end up in a situation of severe information asymmetry, which is the fundamental problem of regulation.

Recall the principle behind benchmarking is that regulators (who might be an elected city commission) cannot rely solely on the information provided by the firm when designing its regulatory framework (including targets and incentives). Since the utility benefits from having private information, regulators would like to avoid such dependence when designing targets for a rate-of-return-based price structure or a price-cap regime (including an X-factor which can specifically incorporate productivity differentials across firms or between the water sector and the economy in general). So, if a benchmarking methodology still requires private information from the firm, one can argue that we have gained little.

Furthermore, mismanaged water utilities are not likely to collect the relevant information needed for effective managerial decision-making—either because the managers lack the skills or because they do not want to document poor performance. In addition, those entrusted with monitoring the utility may not do their jobs: Water Ministries, legislative committees, and mayors (in the case of publicly owned firms) may use utilities for political purposes, and have minimal incentive to collect accurate information regarding performance. Similarly, in the case of privately owned firms, the Board of Directors may have an underdeveloped internal governance system. Managers of both publicly- and privately-owned water utilities may present information in a selective manner to limit the impact of organizations providing oversight. Thus, the problems of data availability and verification are omnipresent. Nevertheless, those concerned with utility performance must begin the process.

After assembling the raw data, the benchmarking team should screen the data carefully to ensure the quality and quantity of information being gathered meets the requirements to allow for a successful outcome. This process is crucial to the study because poor data quality (including the presence of outliers) may lead to biased results. Specifically, attention must be paid to the following aspects when investigating raw data and their quality:

Missing data: Missing data is a common problem for developing countries, which may pose a serious problem for the benchmarking study for the following reasons.

- **Sample size:** Missing data will reduce sample size. As mentioned above, a reasonable sample size is necessary for conducting sophisticated frontier benchmarking models such as Data Envelopment Analysis (DEA) and Stochastic Frontier Analysis (SFA). This issue becomes one of the major obstacles for many developing countries when conducting these kinds of studies. Many countries only have data for a few water utilities covering a short time period (due to previous mismanagement). Missing observations further reduce what is often already a limited sample size, which makes sophisticated studies even more difficult.[4]
- **Nonrandom missing data:** The missing data may be non-randomly distributed. For example, data providers (utilities) could have accidentally forgotten to supply data-randomly distributed missing data; or they could have intentionally not provided the data because the observations might demonstrate poor operations. The nonrandom missing data will influence the statistical inferences because the observed sample will be a truncated sample of the true sample. Even for nonparametric models, deliberately withheld data will influence the benchmarking results because firms with poor performance might be omitted systematically from the sample.

Remedies: There are three ways to address the missing data problem:

(1) Delete the observations with missing values for variables,
(2) Estimate (impute) missing values using existing observations and then use the estimated values in the data analysis, or
(3) Ask the utility to provide the missing data.

As explained above, option (1) will reduce the sample size and option (2) may be inappropriate where the missing data are not random. Therefore, option (3) is recommended since it will reduce the potential problems we mentioned above, although it certainly introduces costs and delays into the process. Having an external (and independent) auditor assist in data collection can improve the process.

---

[4] This observation is based on a data collection initiative undertaken with funding from the Inter-American Development Bank. Maria Luisa Corton and Sanford Berg (2009 forthcoming) "Benchmarking central American water utilities," *Utilities Policy*.

Accuracy and comparability of data files: Before moving into formal statistical analysis, it is helpful to engage in preliminary data checks of the computerized file:

(1) **Reduce data entry errors:** Compare the computerized data set with the original data set to make sure that the data have been entered correctly.
(2) **Prepare summary statistics:** Summary statistics include the mean, median, standard deviation, and minimum and maximum values of the key variables. Based on the summary statistics, analysts need to make sure that the utilities in the sample are really comparable to one another. For example, is utility A really comparable to utility B which is many times larger than A? Exclude the extreme case from the sample to make the sample more homogeneous and comparable.
(3) **Calculate key ratios:** Prepare summary statistics for key ratios such as water delivered/employee and OPEX/water delivered. Pay attention to firms with extremely large or small ratios (those that deviate from the mean by more than 3–4 standard deviations). There might be some data input errors for these firms or they may have different organizational structures (e.g., outsourcing),
(4) **Ensure comparability of data definitions:** Check the definitions of the key variables to make sure that they are really comparable. For example, does the number of employees refer to full time employees, part time employees, or total employees? Is the definition of operating cost similar across firms? In some cases, number of customers may not be a good output variable due to significant differences in service continuity. For instance, let company A's number of customers be 1000 and average service time 4 hours/day. Company B's number of customers is 600 and the average service time is 24 hours/day. An adjusted customer number (reflecting service hours) might be used as an output measure, depending on the structure of the model. However, explicitly including service continuity as an output would be better.
(5) **Recognize unique events or characteristics:** Some special events such as natural disasters may have significant impacts on operating costs and service quality indicators. If the impact is severe, it is better to exclude the observation from the sample in that year. Similarly, a utility located near a water source and one requiring substantial investments in water storage and transport will have different costs. Such differences should be incorporated into the analysis through the inclusion of appropriate variables.

2.2.2 Assemble benchmarking dataset

After investigating the data, the benchmarking team can assemble the benchmarking dataset. However, the sample size may not be large enough for utilizing sophisticated frontier models. Issues include model specification (and degrees of freedom) and ways to expand the sample size.

Model specification and degrees of freedom: For example, when comparing the relative efficiency of 10 utilities in a specific year, the analyst might consider using a translog production frontier with three inputs (specific functional forms are described below). These inputs could be labor, capital, and energy. Due to the translog functional form, there will be 9 regressors since the three independent variables expand into nine variables to take into account interactive effects (as when having a lot of capital affects how additional labor contributes to output). This functional form leaves insufficient degrees of freedom for the estimation. If the number of observations (sample size) is not substantially greater than the number of parameters being estimated, the results will not be robust. To solve this problem, a simple function form should be used (e.g., a log-linear Cobb-Douglas function with fewer parameters to be estimated, like the one used in the Portuguese study described earlier). However, the trade-off is that Cobb-Douglas function involves making more assumptions about underlying production relationships.[5]

Expansion of sample size: Another way around this problem is try to augment the sample size. There are two options available: use the panel data or conduct the international (cross regional) benchmarking. However, as was noted above, international (cross regional) benchmarking is risky due to the potential differences in accounting standards, economic conditions, and geographical environments. Therefore, the benchmarking team should first consider the panel data sample. If the panel data sample is still not large enough, international (cross regional) benchmarking can be considered.

2.2.3 Analyze the data and conduct performance benchmarking study

The process of model specification and technique selection process depends on benchmarking objectives, data availability, and the team's willingness to adopt specific assumptions for each type of model. Hence, the benchmarking team

[5] For now, the technical details are not so important, but the Cobb-Douglas log-linear specification assumes the same production elasticities (regardless of the level of other inputs), the same scale elasticities (again, regardless of the level of other inputs), and unitary elasticities of substitution for all firms.

may need to draw upon professional consultants or specialists at research institutions before moving to more sophisticated models. Two broad approaches to quantifying performance are outlined here:

Data envelopment analysis: DEA is a method in which linear programming techniques are applied to a selected set of variables to calculate an efficiency coefficient for each water utility. The framework has been adapted from multi-input, multi-output production functions and applied in many industries. We have seen how the production function is a basic concept in economics used to determine how much output can be produced with a given basket of inputs. A cost function relates total cost to input use and input prices. DEA develops a function whose form is determined by the most efficient producers. This method differs from the Ordinary Least Squares (OLS) statistical technique that bases comparisons relative to an average producer. DEA benchmarks firms only against the best producers (so it is a frontier analysis). It can be characterized as an extreme point method that assumes that if a firm can produce a certain level of output utilizing specific input levels, another firm of equal scale should be capable of doing the same. The most efficient producers can form a "composite producer," allowing an efficient solution for every level of input or output. Where there is no actual corresponding firm, "virtual producers" are identified to make comparisons. DEA is a non-parametric approach to evaluating performance: functional forms are not specified, although returns to scale can be incorporated into the model.

DEA has been used for both production and cost data. Utilizing the selected variables, such as unit cost and output, DEA software searches for the points with the lowest unit cost for any given output, connecting those points to form the efficiency frontier. Any company not on the frontier is considered inefficient. A numerical coefficient is given to each firm, defining its relative efficiency. Different variables that could be used to establish the efficiency frontier are: number of employees, service quality, environmental safety, and fuel consumption. An early survey of studies of electricity distribution companies identified more than thirty DEA analyses—indicating widespread application of this technique to that network industry.[6] A number of studies using this technique have been published for water utilities. The main advantage to this method is its ability to accommodate a multiplicity of inputs and outputs. It is also useful because it takes into consideration returns

[6] T.J. Jamasb and M.G. Pollitt (2001). "Benchmarking and regulation: international electricity experience." *Utilities Policy*, **9**(3), 107–130.

to scale in calculating efficiency, allowing for the concept of increasing or decreasing efficiency based on size and output levels. A drawback of this technique is that model specification and inclusion/exclusion of variables can affect the results.

Statistical analysis: Regression analysis seeks to derive a relationship between firm performance and market conditions and characteristics of the production processes. Particular specifications (and/or data transformations) can capture non-linearities. Statistical analysis of production relationships can isolate the effects of specific conditions or input levels on the level of output—so the roles of multiple independent variables can be determined. Data from the firms being compared can then be used to calculate expected output, given the model coefficients and values of variables for each firm. The process is similar for a cost function. A major advantage of this method is its ability to reveal information about cost structures and distinguish between different variables' roles in determining cost. Predicted versus actual cost can provide a measure of relative performance. The sensitivity of these results to changes in model specification or data points can then be examined to provide the policy-maker with a framework for evaluating firms.

Econometric analyses have two major disadvantages. First, a large data set is necessary for reliable results. Obtaining the number of observations needed to derive an efficient and unbiased estimate of cost (or production) structures can often prove to be a difficult task. The second disadvantage is a statistical one. The regression results are sensitive to model specification (for example, a linear vs. a non-linear functional form or the exclusion of particular explanatory (independent) variables. In addition, for some models, the interpretation of the error term becomes important (as with Stochastic Frontier Analysis, a more complicated approach than Ordinary Least Squares—OLS). Nevertheless, these techniques are widely used to analyze other network industries.

The early studies tended to utilize OLS to estimate cost functions for firms. Due to data limitations, most of these studies were cross-sectional in nature. Besides using data from only a single year, researchers tended to utilize data from England and Wales or from the United States. These academic studies often focused on the relative performance of private vs. publicly-owned water and sewerage utilities. In addition, they investigated the extent of scale economies and economies of joint production (providing both water and sewerage services). In some cases, they considered the impacts of residential vs. industrial/commercial customers.

As data from Brazil, Peru, and other emerging nations became available, additional country studies were published—often using more advanced econometric (parametric) or non-parametric data analysis techniques. Studies of utilities in France, Italy, and other nations began to appear in the academic literature. Techniques associated with Stochastic Frontier Analysis began to be applied to both production functions and cost functions. Panel data facilitated the incorporation of customer density, topology, and other variables.

In addition, a number of cross country studies have been conducted by researchers at the World Bank.[7] These studies often characterize production processes in a region and to draw conclusions about the impacts of different institutional features (such as the presence of good governance procedures, the regulatory environment, and public/private ownership). Furthermore, quality of service has begun to be incorporated in empirical studies, through the use of hedonic output measures (adjusting the output for quality) or as separate outputs altogether (in multi-output models).

For this sub-step, we have already assumed that the benchmarking team is conducting a Metric Performance Benchmarking Study (rather than one involving engineering models, process benchmarking, or customer service benchmarking). The team still needs to select from among the four types of techniques, based on performance benchmarking objectives and data availability. We will briefly discuss the data requirements, advantages, and disadvantages for each type. The discussion of specific sub-type techniques (Step 3) are also incorporated in this discussion. More comprehensive discussions of the techniques are presented in the Appendix 3: Technical Features of Benchmarking Methodologies.

Figure 4 depicts the various analytic techniques available for analyzing production and cost relationships.

Benchmarking specialists can be expected to master (or at least understand) the technical features of the various approaches to benchmarking. Those comfortable with the use of various quantitative methods will find the short summary of methodologies in Appendix 3 fairly straight-forward. The key lesson for practitioners is that each method has strengths and limitations. The choice of methodology depends on data availability and the issues under investigation. Performance scores that are based on the different methodologies are only as accurate as the underlying models. Furthermore, the application of these approaches requires substantial technical skill.

---

[7] A. Estache and M. Rossi (2002). "How different is the efficiency of public and private water companies in Asia?" *The World Bank Economic Review* **16**(1), 139–148. C. Nauges and C. van den Berg (2008). "Economies of density, scale and scope in the water supply and sewerage sector: a study of four developing and transition economies," *Journal of Regulatory Economics*, **34**, 144–163.

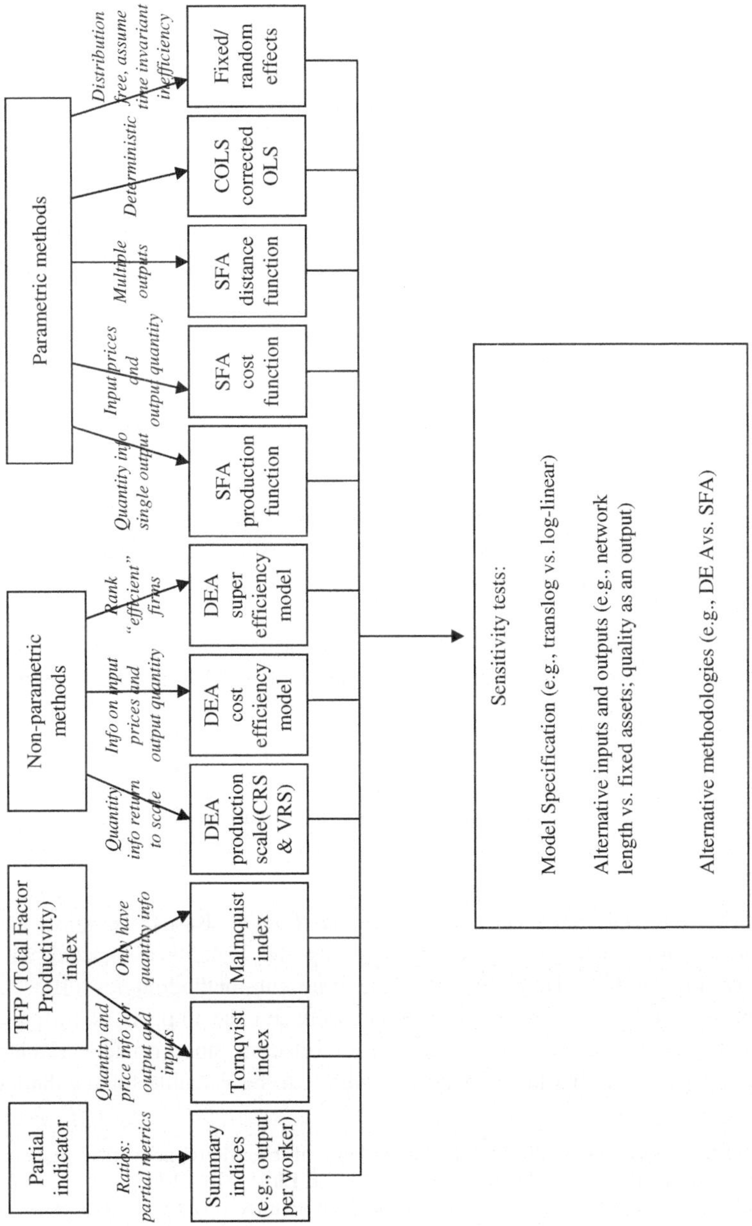

Figure 4. Step 3: Utilize Specific Analytic Techniques

2.3 Utilize Specific Analytic Techniques (Step 3)

In many cases, research raises more questions than it answers. Public policy-makers would like empirical studies to answer questions that have decision-relevance. The publications and studies summarized in the Technical Appendices have primarily involved academic researchers, sometimes in the context of rate cases—or as a retrospective look at how methodologies were used by regulators. Consulting firms have entered the fray, as well. Water utility managers are likely to fear the misuse of empirical studies: they tend to prefer process benchmarking studies that can identify changes in specific practices that would improve performance.

Skeptics of sophisticated models might argue:

If you torture the data enough, it will confess.

This concern has some basis. When substantial sums of money (or managerial careers) are at risk, parties with divergent interests might acquire "hired guns" who can obtain "desired" results. Generally, consultants and academics have reputations to maintain, which limits the likelihood that the results are completely dictated by the desires of the study sponsors. Nevertheless, the robustness of the results to different quantitative methods and model specifications is an important consideration for establishing credibility of the results.

Respected researcher Graham Shuttleworth[8] argues that more elementary total factor productivity approaches may be superior to DEA and SFA approaches to benchmarking:

> "Benchmarking may still have a role to play in regulation, as long as regulators recognize that the residual merely measures the extent to which the model has failed to explain costs, and not the extent to which companies are inefficient. Stable and predictable regulatory techniques have to rely on other forms of evidence." (p. 317)

However, in the same issue of *Utilities Policy,* Burns, Jenkins, and Riechmann[9] are much more positive regarding the benefits associated with benchmarking studies. They note that the different methodologies differ in assumptions, in their ability to discriminate between the impacts of different explanatory variables, and the importance of selecting appropriate variables. However, they argue that benchmarking studies can be valuable for regulators.

---

[8] Graham Shuttleworth (2005). "Benchmarking of electricity networks: practical problems with its use for regulation," *Utilities Policy*, **13**(4), 310–317.

[9] Phil Burns, Cloda Jenkins, and Christoph Riechmann (2005). "The role of benchmarking for yardstick competition," *Utilities Policy*, **13**(4), 302–309.

Shuttleworth also finds that benchmarking can help in the following cases:

- "To decide which companies out of a large sample deserve closer (and more objective) examination, so that scarce investigative resources are allocated efficiently;
- To appraise individual cost items, if their cost drivers are a matter of consensus;
- As a preliminary step in the detailed investigation of each company's costs, aimed at identifying decisions that should be investigated further for 'imprudence'." (p. 317)

Overall, the process of model specification and technique selection process is not easy. The appropriate approach depends on benchmarking objectives, data availability, team capabilities (and resources), and the team's willingness to adopt specific assumptions within each type of model. Hence, the benchmarking team may need to hire professional consultants or confer with specialists at research institutions before moving to the more sophisticated model. The detailed discussion of Step 2 in Appendix 3 outlines the strengths and limitations of the different analytic techniques:

Partial indicators

(1) Summary indices (Core indicators)

Total factor productivity

(2) Tornqvist index
(3) Malmquist index

Nonparametric

(4) DEA Production scale
(5) DEA Cost efficiency
(6) DEA Super efficiency

Parametric

(7) SFA Production function
(8) SFA Cost function
(9) SFA Distance function
(10) Corrected ordinary least Ssquares (COLS)
(11) Fixed and random effects models

These methods are (basically) arrayed in terms of increasing technical quantitative skills required for implementing the different approaches. Availability of software and specialized skills can be as important as data availability in determining the types of models utilized. Summary indices allow analysts to review trends of core indictors over time. Total factor productivity indices provide more comprehensive characterizations of trends over time and of relative performance across a set of water utilities. However, these indices (and those emerging from parametric and nonparametric methods) may not incorporate a number of important dimensions of performance noted above (e.g., financial sustainability or some aspects of customer service or water quality) due to difficulties in incorporating these factors into DEA, SFA, or COLS models. Thus, more sophisticated techniques are not necessarily the most useful in the context of benchmarking. Sometimes simple partial indicators can provide useful comparisons that are intuitive and easy to interpret and communicate. In addition, credible performance comparisons require that the results be communicated in a manner convincing to stakeholders. A very technical analysis that is understood only by a select few is unlikely to persuade decision-makers. Thus, the robustness of quantitative benchmarking analyses is crucial—which brings us to the role of sensitivity tests.

2.4 Sensitivity Tests (Step 4)

In making performance comparisons through benchmarking analyses, the analyst is interested in obtaining a measure of firms' relative efficiency. Such information can be used to develop X-factors in price cap regulation, to reward (or punish) companies. Or the regulator might want to publish the rankings or efficiency scores to provide the public with information, putting pressure on managers of poor performing utilities to improve the performance of their firms. In both cases, the accuracy and robustness of inefficiency estimates are very important because they may have significant financial or social impacts. In particular, if the estimated inefficiency scores or rankings are sensitive to the benchmarking method, a more detailed analysis is required to justify the adopted model. Tests for mutual consistency are becoming standard.

2.4.1 Robustness of results

In most cases there is no "ideal" model among the set of potential models. Issues include model specification (cost vs. production function and the choice of specific functional form), alternative specification of inputs or outputs (e.g., network length vs. fixed assets—from the Balance Sheet as an input), assumptions about error terms, and alternative methodologies (e.g., DEA vs. SFA). Following the work of others, we suggest three levels of sensitivity tests to determine confidence in the performance of comparisons, which are displayed in Figure 5.

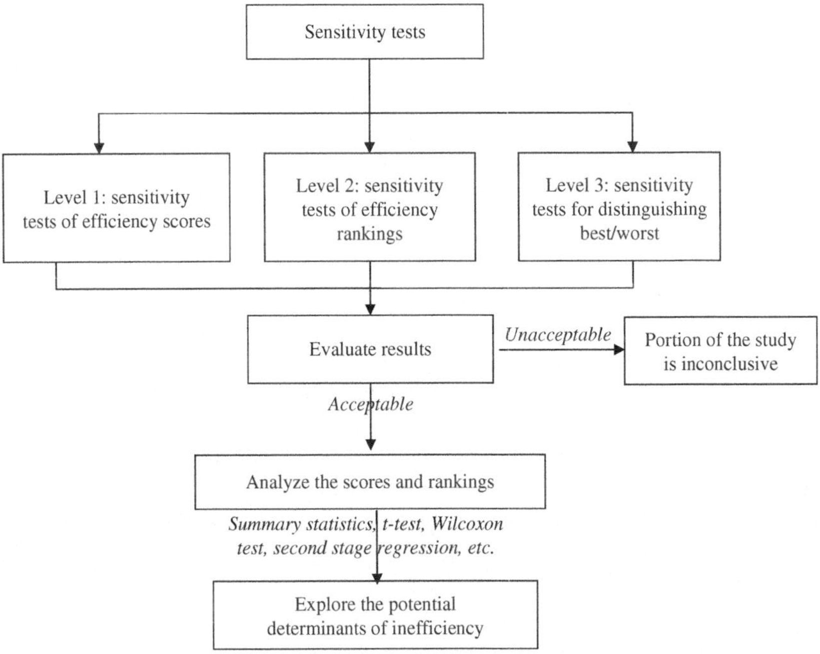

Figure 5. Step 4: Sensitivity Tests

To check for the robustness of performance rankings, researchers have begun to compare results from different methodologies: using correlation matrices or verifying whether different models identified the same set of utilities as the most efficient and least efficient firms. Clearly, if efficiency scores are to have any use for managerial incentives or as elements in regulatory mechanisms, stakeholders need to be confident that the scores reflect reality, and are not just artifacts of model specification, sample selection, treatment of outliers, or other steps in the analytic process. Thus, analysts are performing sensitivity tests.[10]

[10] Berg and Lin (2008). "Consistency in performance rankings: the Peru water sector," *Applied Economics*, 2008, 793–805.

Benchmarking methodologies can produce dramatically different results under similar circumstances. A quick review of the benchmarking literature reveals that conclusions (such as performance rankings or scores) differ considerably, depending on the

- variables chosen,
- the particular methodology applied,
- the interval of time considered, and
- other factors that need to be determined for a benchmarking study.

This point applies to all the methodologies used: from simple ratio analyses to sophisticated quantitative techniques (stochastic frontier analysis, Data Envelopment Analysis, and distance functions). This observation raises questions about how reliable a benchmarking process can be, especially when we realize that performance comparisons affect the economic foundations and financial sustainability of a company that is usually the sole provider of WSS service in a particular geographic area. This important issue brings into question the use of complicated mathematical techniques that are sensitive to manipulation.

Bauer *et al.* (1998)[11] propose a series of criteria that can be used to evaluate whether the results obtained from different methods are ''mutually consistent,'' that is, if they lead to comparable inefficiency scores and rankings. We will discuss these criteria in detail in the next step. Coelli and Perelman (1999)[12] have suggested combining the results from alternative modeling exercises by using the geometric means of the performance scores for each data point in order to reduce potential bias. Thus, attention should be paid to the sensitivity of the results to the following elements:

- Model Specification (e.g., translog vs. log-linear)
- Alternative inputs/outputs (e.g., network length vs. fixed asset)
- Alternative methodologies (e.g., DEA vs. SFA)

Figure 5 outlines the steps in terms of a flow diagram.

---

[11] P. Bauer, A. Berger, G. Ferrier, and D. Humpherey (1998). "Consistency conditions for regulatory analysis of financial institutions: a comparison of frontier efficiency methods," *Journal of Economics and Business*, **50**, 85–114.

[12] Coelli, Tim and Perelman, Sergio (1999). "A comparison of parametric and non-parametric distance functions: with application to European railways," *European Journal of Operational Research*, **117**(2), 326–339, September.

2.4.2 Three levels of sensitivity tests

A problem faced by regulators willing to apply frontier studies is the variety of alternative model specifications and methodologies available. The problem becomes even more serious if different methods generate substantially different results. Bauer etal.'s proposed sets of consistency conditions can be used to analyze results given different model specifications. Different methods should provide consistent efficiency levels and rankings. Their approach as involves three levels of consistency tests summarized below.

Level 1: Sensitivity tests of efficiency scores: A Pearson correlation matrix can be employed to check the correlation of efficiency scores between pairs of techniques. Furthermore, Kruskal-Wallis nonparametric test can be used to test the null hypothesis that different techniques generate the same distribution of efficiency scores.

Level 2: Sensitivity tests of efficiency ranking: If the efficiency scores are not consistent across the different methods, it is still possible that these approaches generate similar rankings of firms by their efficiency score. A clear ranking can help a regulator determine the X factor to be used in setting prices for the firms in the sector. Thus, nonparametric Spearman's ranking correlation matrix can be used to check the correlation of rankings between pairs of techniques.

Level 3: Sensitivity tests for distinguishing best/worst: If the consistency in efficiency level and rankings was not met, it is still possible that these approaches can identify the best and worst performers, which can be especially helpful for rewarding the best performers and punishing the worst performers. Therefore, the regulator can compare rankings yielded under the different techniques and summarize the extent to which results overlap, allowing anlysts to identify the best and worst performers.

After the three levels of tests, the regulator should have a good sense of the consistency of different methods. The regulator should pay more attention to firms that have similar efficiency scores/ranking under different approaches (for instance, a firm which is recognized as one of the worst performers by 3 different techniques). If the results are close to each other, we can calculate the geometric means of the efficiency scores for each firm to get a "comprehensive" efficiency measure. If there is substantial variation in all three level tests, the findings would be considered inconclusive; requiring a more detailed analysis to explore problems with the adopted models. This step is more "art" than "science" at present, but there is certainly a need for considering this issue in a systematic manner.

2.4.3 Analyze the scores and rankings and explore the potential determinants of inefficiency

If the results pass the sensitivity test hurdle, the benchmarking team can start to analyze scores and rankings and explore the potential determinants of inefficiencies across firms and over time. The utilities can be divided into different groups by various factors, such as regions, population density, regulatory environment, ownership structure, and asset vintage to compare the efficiency scores. Second stage regressions (OLS or Tobit) with the efficiency score as the dependent variable can also be used to test the partial effects of these external factors on firm efficiency. Firms should not be ranked as poor performers if they operate under conditions that differ from those of the other firms. As noted earlier, density, geographic topology, distance from raw water sources, and political constraints on prices (affecting the financial sustainability of operations) all affect relative performance. Given the complexity of benchmarking studies, it is useful to seek public comments that provide feedback, both augmenting the effort to analyze performance and promoting transparency and the legitimacy of the process.

Exploring the potential determinants of inefficiency might require public workshops. The process will certainly require participation of stakeholders. As has been noted before, utility managers have access to far more detailed information regarding production processes, opportunities for improvements, and costs associated with different strategies for cost containment, quality of service improvements, and other activities. Thus, performance rankings should not be viewed as ends in themselves. Rather, they are catalysts for promoting critical thinking about the sources of inefficiency. Next we identify steps in this process.

2.5 Develop Policy Implications (Step 5)

The reason for resolving the methodological issues is to enable the analyst (and those using the study) to have confidence in the performance evaluation. Policy implications depend on how well the methodologies meet the principal *criteria for choosing* a set of performance measures: they have to be comprehensible, comprehensive, useable, and timely. We return to these issues later in this volume, in the context of incentives for strong performance. Consider Figure 6.

2.5.1 Explore the potential determinants of inefficiency

Through the summary statistics and second stage regressions, the benchmarking team can identify the potential determinants of firm inefficiency, which might

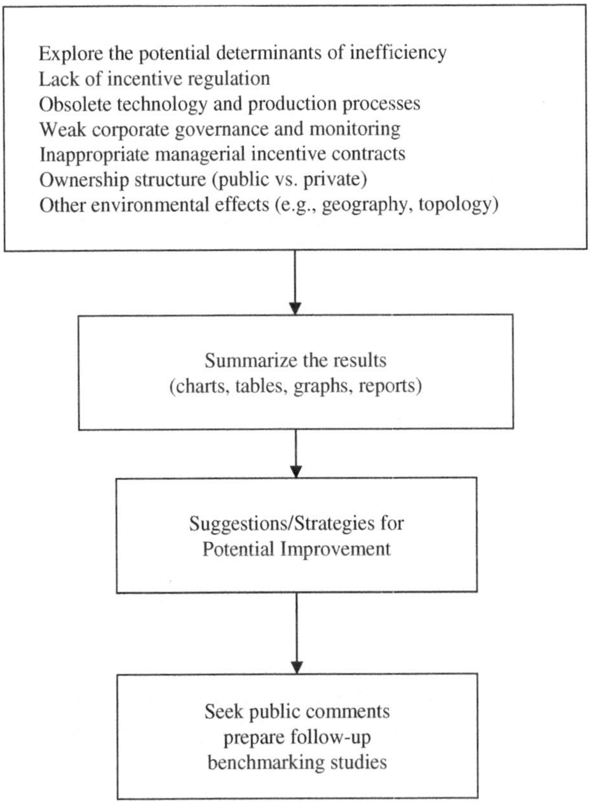

Figure 6. Step 5: Develop Policy Implications

include but are not limited to the following factors:

- Lack of incentive regulation
- Obsolete technology and production processes
- Weak corporate governance and monitoring
- Inappropriate managerial incentive contracts
- Ownership structure (public vs. private)
- Other environmental effects (e.g., geography, topology)

A detailed discussion about the definition, measurement, and potential impacts of these factors is presented in Appendix 1: Variable Definitions and Explanations.

2.5.2 Summarize the results: formats for presenting comparisons

Given the various audiences for the results of benchmarking studies, care must be taken when presenting rankings. The press filters and highlights reports, using executive summaries and interviews. Although technical reports are not amenable to sound bites, most newspaper and television journalists seek the clear message that emerges from a benchmarking study. However, some seek sensational factoids that support their own ideological predilections and some lack the expertise to interpret technical studies. Similarly, the general public is not well-positioned to evaluate conflicting claims. Long before a benchmark comparison is released, the responsible agency should be engaging in an information-dissemination campaign, informing political leaders and the press about the purpose of the forthcoming report. NGOs and formal citizen advisory committees that can be established by regulatory commissions provide opportunities for input and feedback for citizens.

Any classification scheme can be misapplied. If the user does not have confidence in the scores or the rankings, then the studies will not be applied in the evaluation process. Some formats are described below to illustrate how studies have tried to present information in a useful manner.

Tables: The simplest ranking scheme involves a table that ranks firms from strongest to weakest. The tables can group the water utilities into categories, directing attention to the firms that are most efficient and to those with the weakest performance. In all likelihood, the order of ranking and the "scores" will not be precise, so grouping allows the performance evaluator to acknowledge that no methodology is perfect. Reporting a score of 87.352 involves misplaced concreteness. No score is likely to have so many significant digits, and even if one model did yield such an indicator, other methodologies would yield other "numbers". In partial response to this issue, in Peru, SUNASS gives each municipal water utility a "grade": for example, a C+ or A. This format provides an intuitive and easy-to-grasp measure of relative performance. In addition, it is possible for all the utilities to earn an A grade.

Coefficients (and scores) from DEA and SFA studies: We have seen how quantitative studies can be very complicated, and the results may not appear to be very intuitive. For specialists, drawing conclusions about economies of scale or the association between service quality and costs involves careful specification and sensitivity tests. Presenting scores to non-specialists (based on the results of such studies) presents a challenge for econometricians and other technicians. This area is one warranting much more attention, since different audiences require different types of reports. For example, instead of focusing on the top five and the bottom five performers (out of a set of twenty), the report might only consider the bottom four if there is a significant gap between the efficiency scores of the 16th and 17th ranked firms and the scores for the 15th and 16th are very close to one another. Or the report might highlight the top six if those scores cluster together. Officials need to use common sense when presenting information that affects managerial careers and public perceptions.

Fingerprint charts or spider webs: Another format allows an individual utility to be compared with another firm and/or with the average of a set of peer utilities. Peter Stahre and Jan Adamsson (2002) utilize six indicators:

- **Customer satisfaction**—Performance indicators and measuring methods reflecting customer expectations and their appraisal of water service.
- **Quality**—Quality-related performance indicators complementing the economic indicators and the customer satisfaction indicators.
- **Availability**—Performance indicators describing the reliability of the entire system's operations.
- **Environment**—Performance indicators illustrating the utilities' environmental achievements.
- **Organization/Personnel**—Performance indicators describing efficiency and the relationship between "in-house work" and external services.
- **Economy**—Performance indicators comparing costs on an overall level.

Figure 7 (taken from their study) illustrates one format used to compare the distribution network in the city of Malmo in relation to the calculated average situation in the Six-Cities group[13]:

> "The general principle of the diagram is that the longer the "beams", the worse are the conditions. Obviously, this is not the case for describing the age structure of the pipe network, which is marked in a deviant way. The fingerprint model is intended for benchmarking water distribution networks in different cities. For this purpose the information must first be normalised. The dotted line in the diagram (= 100%) represents the mean values of the respective parameters within the compared group of cities. The "fingerprint" represents the situation in one city and indicates how this city is positioned in relation to the calculated mean value for the group. The information in the diagram is thus given in percentage of the average."

[13] Paul Stahare and Jan Adamsson (2002). *Performance Benchmarking – A Powerful Management Tool for Water and Wastewater Utilities*, Suecia.

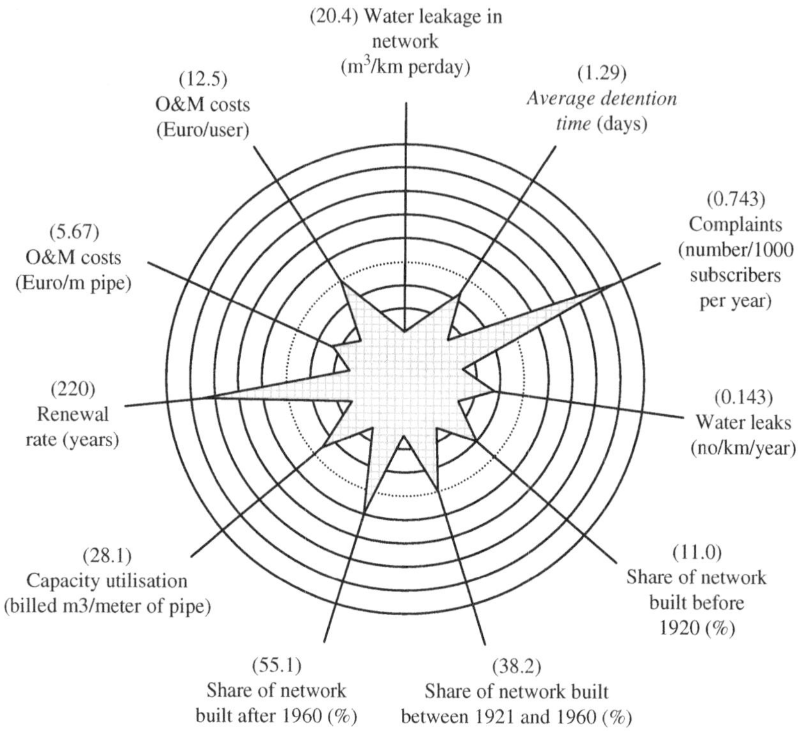

Figure 7. Characteristics of the Water Distribution System of Malmo

The format allows one to see key features of the network in a relatively compact way. "Every increment is 20%. The average in the Six-Cities group is indicated by the dotted line (= 100%). The average in absolute numbers is given in parenthesis for every parameter." In general, the tighter the "star", the better, except for the "share of network" elements which help the analyst understand the relative age of the networks. Other factors, such as soil conditions or topology, could be incorporated into the comparison. The fingerprint model is particularly useful for characterizing multidimensional "core performance indicators" of a pipe network. However, it is probably more useful for professionals than for the average citizen.

Radar charts represent a similar concept used by regulators in Canada to compare local utilities against efficiency goals. Utilities are given percentage rankings for specified goals. The resulting diagram, much like the fingerprint chart, readily identifies areas that need to be improved upon. Problem areas are identified in the charts by low percentage scores).[14]

OPEX and CAPEX Tradeoffs: OFWAT has conducted quantitative studies of relative performance for water utilities. The agency funded statistical studies of operating expenditures (OPEX) as a function of cost-drivers. Different categories of cost were related to factors that influenced the level that cost component, such as network length and volume delivered. OFWAT published the studies (and estimated coefficients) and ranked the firms in terms of whether costs were far above (E), above (D), about equal to (C), below (B) or far below (A) "expected costs". The top grade of A implied that operating costs were far less than what the model predicted, given network length and other operating conditions. Thus, the ranking was based on the difference between actual vs. the estimated costs, where the difference was derived from statistical cost studies. It must be noted, however, that low operating costs may due to excessively reducing or deferring maintenance. A firm should not be rewarded if costs are being controlled by not spending on maintaining the water system. Such managerial policies will only necessitate higher levels of spending at a later date due to rapid deterioration of the system.

For capital expenditures (CAPEX), OFWAT used a different methodology for ranking firms from A to E. Utilities were asked to provide estimates of the cost of specific capital projects, and the utility estimates were compared with those of independent engineers (and estimates by the other utilities). So this component takes on an engineering ("model firm") approach to establishing performance scores regarding CAPEX. In recognition of the substitution possibilities between capital and operating expenditures, OFWAT presented the data in a matrix (Figure 8) that identified an AA firm as "clearly outstanding" and an EE firm as quite weak. A firm that scored A on OPEX and E on CAPEX was still viewed as having costs that were overall "lower than expected." The performance rankings were used to determine X factors, with the utilities exhibiting highest performance given lower productivity off-sets in the water utility price cap regime. For more on implementing price caps, see studies by Ashton (2000)[15] and Cubbin (2005).[16] These authors have different viewpoints regarding OFWAT's use of benchmarking methodologies.

---

[14] D. Main, L. Ng., and A. North (2008). "The Canadian National Water and Wastewater Benchmarking Initiative: Using Process to Drive Improvement—Strategic Management of Water in Urban Areas," in Caberra and Pardo op. cit.

[15] J.K. Ashton (2000). "Cost efficiency in the UK's water and sewerage industry," *Applied Economics Letters*, **7**, 455–458.

[16] Cubbin, J. (2005). Efficiency in the water industry, *Utilities Policy*, **14**, 289–293.

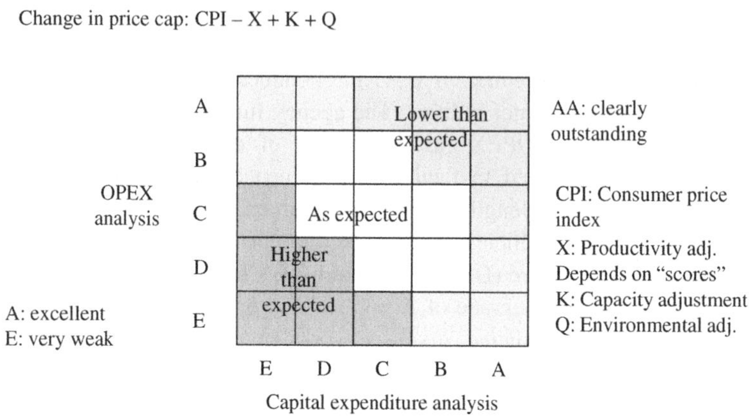

Figure 8. OFWAT Trade-offs: Performance Benchmarking

2.5.3 Suggestions/Strategies for potential improvement

After conducting the analysis and checking the sensitivity of the results to alternative specifications and methodologies, the benchmarking team can put forward suggestions or strategies for potential improvements, which might include but are not limited to the following strategies:

- Adopt incentive regulation
- Update technology and production processes
- Strengthen corporate governance and monitoring
- Improve managerial incentive contracts
- Promote privatization or private participation
- Publish the ranking to stimulate the poor performer
- Use the efficiency score to decide the X factor under a price cap regime

The role of information as a catalyst for change cannot be underestimated. Most citizens are unaware of how their local monopoly water supplier stacks up against comparable utilities. Similarly, managers may not be fully aware of their own performance. It is psychologically comforting to attribute good developments to our own skill and bad outcomes to things outside our control. Reality-therapy can be jolting, but living in a fantasy is not healthy—for individuals or organizations. Information on relative performance sets in motion corrective processes that ultimately benefit stakeholders.

2.5.4 Follow-up benchmarking studies

Good studies often raise as many questions as they answer. Nevertheless, decision-makers need to avoid "paralysis by analysis." No study is perfect, but studies can be used to place the burden of proof on parties arguing that the analysis is incomplete or incorrect. It is clear that information asymmetries pervade the regulatory process. Benchmarking provides a stimulus for additional studies by stakeholders. Of course, a trained (but unethical) professional can obtain desired results by dropping observations, changing time periods, or experimenting with alternative specifications. This possibility explains why follow-up studies are necessary, so that stakeholders gain confidence in the process.

Rossi and Russier[17] prepared an excellent overview of benchmarking methodologies and their application in the regulatory process. They note the cumulative nature of benchmarking studies:

> ...regulators can seek the involvement of the firms in the benchmarking process to ensure that the data on which the analysis is based are reliable and that the results are comprehensible and justifiable. Yardstick competition would then result in a 'learning by doing' iterative process in which both firms and regulators learn while playing the game. (p. 90)

2.6 Recent Institutional Developments

Developments over the past decade in quantitative techniques and pressures for sector reform have stimulated interest in identifying and understanding the factors that can contribute to WSS network expansion, improved service quality, and cost containment in the sector. Policymakers in Latin America, Asia, and Africa have begun to collect data that can serve as the basis for performance comparisons—creating yardsticks that help decision-makers identify weak and strong performers. Water utility managers in Europe and North America have developed performance indicators that served as the basis for later work, although highly sophisticated data collection systems are costly. Utility managers, water associations, regulators, and other groups have begun to undertake statistical analyses of water systems—over time, across geographic regions, and across countries.

In addition, a number of other initiatives are bringing together networks of regulators and utility managers. The Asociación de Entes Reguladores de Agua

[17] Martin Antonio Rossi and Christian Alejandro Ruzzier (2000). "On the regulatory application of efficiency," *Utilities Policy*, **9**(2), 81–92.

Potable y Saneamiento de las Américas (ADERASA—Water Regulators of the Americas) received World Bank seed money and established a Benchmarking Task Force. The Water Utility Partnership for Capacity Building in Africa (WUP) has promoted mechanisms for sharing data, as have SEAWUN (a group of water associations in South East Asia), water associations in other regions, and SNIS (housed in the Brazilian Ministry of Cities).

To further facilitate the exchange of information, the World Bank and WSP have supported the creation of a Water and Sanitation International Benchmarking Network (IBNET) to serve as a clearinghouse that brings together data from a number of utilities in a single site with common data format. The IBNET initiative represents a significant step in developing a set of core indicators and data definitions, a data capture system, and mechanisms for sharing information on water utilities. Table 1 summarizes some of the material developed in the Appendix 4 which describes benchmarking activities around the world.

The World Bank has recognized the value of creating resources for policymakers. For example, the Water Tool Kits provide useful information on a number of topics, including the design of contractual arrangements that can improve efficiency. Another resource is the Body of Knowledge of Infrastructure Regulation at www.regulationbodyofknowledge.org. This site surveys key issues in utility regulation and management, provides over 500 PDFs, and includes answers to Frequently Asked Questions.

Despite the growing number of decision-support tools, benchmarking is still a difficult process. As has been noted a number of times, every firm is different. Even when two utilities provide the same service mix in similar areas, the operating environment is seldom the same. There is always a particular input, geographic feature, or specific technological consideration that differs from one firm to the other, raising doubts about the possibility of a fair comparison between the two WSS utilities. Inherited infrastructure is one of these features, since it is rare to see firms starting from zero and building entire networks and facilities as a Greenfield activity. In the case of privatized WSS utilities, they have inherited fixed assets already designed and installed years or decades ago. Publicly-owned utilities have often received soft loans or grants in the past, leading to networks that reflected past political priorities. Thus, these regional collaborations provide important opportunities for evaluating data and the studies that follow naturally from data collection.

In short, the five steps described in this section provide a framework for managers, regulators, and academicians to contribute to our understanding of infrastructure performance. No study will be perfect, but that should not halt attempts to better understand the relative performance of water utilities.

Table 1. Overview of Regional Initiatives in Water Benchmarking

Region	Organizations	Established	Participants	Activities	URL web
Africa	South African Association of Water Utilities (SAAWU)	April 2001	Water service providers	Data collection, education, advocacy	www.saawu.co.za
	Water Utility Partnership (WUP)	September 2000	Water utilities	Data collection and analysis	www.wupafrica.org
Asia	Southeast Asian Water Utilities Network (SEAWUN)	August 2002	Water associations	Data analysis, education	www.seawun.org
South America	Association of Water and Sanitation Regulatory Entities of the Americas (ADERASA)	August 2003	Regulatory commissions	Data exchange, collaboration	www.aderasa.org
Middle East	Arab Countries Water Utilities Association	2007	Water utilities	Collaboration	www.acwua.org
Central Europe	EU-funded initiatives				Waterbench 1
Central Asia					Institute of Urban Economics (Moscow) http://www.euwi.net
OECD	EU Water Initiative (EUWI)	2000	Water utilities, municipalities	Data collection, analysis, and collaboration	
	American Water Works Association (AWWA)	1870	Water supply professionals	Education, advocacy, communication	www.awwa.org

Chapter 3 Overview of Metric Benchmarking Concepts

To provide additional intuition behind production functions and cost functions, we need to dig a bit deeper into technical features of production and cost models. This section should be viewed as tutorial on the conceptual framework that economists utilize in analyzing productivity and efficiency. First, let us consider examples that illustrate key concepts.

3.1 Production Concepts

The relationship between inputs and outputs presented here is highly stylized. However, the principles have wide applicability, and both regulators and managers can relate to the underlying principles. We have seen how a single input can be related to volume of water produced. With multiple inputs, the "recipe" linking different input combinations to the output level is a bit more complicated but still has clear intuitive meaning.

Consider the particular production function used earlier: $Q = 10\,L - 0.2\,L^2$. In this simplistic case, output increases with the variable input, labor. However, the negative term indicates that output does not just rise proportional to Q: as L increases, output increases as a decreasing rate. The impact of additional labor on output depends on what range of outputs are under consideration. We will see this point shortly. Furthermore, the role of fixed inputs is important. If N is network size, a more comprehensive production function (the "recipe" linking output to inputs) might be

$$Q = N(10\,L - 0.2\,L^2)$$

Now, we can compare two utilities: one with a network size of "1" and another twice as large, "2".

Quantity (Q)
for a "small" utility (N = 1)

Q	L	N
0	0	1
9.8	1	1
19.2	2	1
28.2	3	1
36.8	4	1
45.0	5	1

Quantity (Q)
for a "large" utility (N = 2)

Q	L	N
0	0	2
19.6	1	2
38.4	2	2
56.4	3	2
73.6	4	2
90.0	5	2

Figure 1 depicts a graphical representation of the implications of larger networks and increased amount of Labor (L).

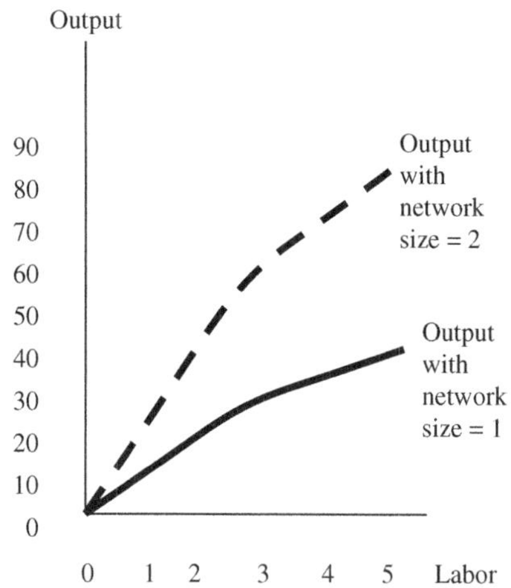

Figure 1. Production Relationship: Small and Large Networks

Note that 4 workers and a network of size 1 yields about the same volume of water as 2 workers operating a network of size 2. Thus, we can see that functions can be specified that contain multiple inputs: here, both network size (N), and Labor (L). More complicated functions could incorporate additional outputs (like number of customers and quality) and inputs (like energy, chemicals for treating water, and the bulk water itself).

For now, we will focus on the two inputs, one output case, where similar relationships could be derived for cost. If we specified the price per worker (per year) and the annualized price of the network, we could determine the cost of producing different levels of output for each of the network sizes. In such situations, the fixed cost (annual cost of the network) would depend on the size of the network; network expansion would be viewed as a long term investment. In this simplistic example, for a fixed network size additional labor (a variable input) could be hired to produce additional output.

The point is that if managers understand the production technology and are attempting to minimize the costs of producing particular levels of output, they will make appropriate input choices (given the input prices) in the short run. In the long run, network expansion reflects adjustments to opportunities to meet the needs of a larger number of customers. Of course, the financial sustainability of the utility will depend on the price charged for water service and access to financial capital. Access to investment funds depends, in turn, on risk perceptions (and/or donor priorities), and projected future cash flows from projects: together, these factors determine the expected net present value of associated net cash flows.

A number of concepts can be introduced and defined:

- *Diminishing returns* applies to the short run because capacity is considered to be fixed at a given point in time, and only variable inputs are being increased. Diminishing marginal returns to inputs explains why short-run marginal cost rises. Of course, for a water network, if there is adequate pipe and pumping capacity, the role of short run diminishing returns may be minimal, so long as the water source is adequate.
- *Returns to scale* is a long-run concept. Increasing returns to scale implies that a 10% increase in all inputs will increase output by more than 10%. Thus, long run average cost (LRAC) is falling over this output range (where LRAC = TC/Q, Total Cost divided by total output). Note that "spreading" overhead over a larger volume of output in the short run should not be labeled returns to scale, since the scale (capacity) of the utility is unchanged.
- *Fragmented industries* have offered opportunities for managers to consolidate production, often via sharing key inputs, such as technical engineering skills or the services of firms checking compliance for

meeting microbiological standards. In the water utility industry, geographic distances and hydrology may reduce some savings from consolidation. However, the provision of specialized services and management skills can be facilitated by more formal collaboration—so long as the jurisdictional lines separating entities can be bridged.

- *Returns to scope* is another long-run concept related to a multi-product utility, for example, a utility supplying both water and sewerage services. Economies of scope refer to whether the addition of entire product lines is efficient.
- *Returns to density* indicate the role of spatial considerations in determining costs. Serving a rural area will tend to be more expensive than serving an urban area, at least in terms of the amount of pipe serving a particular number of customers.
- *Vertical integration* reflects the extent to which a utility handles all the stages of production, from water withdrawal, to treatment, and to delivery to final customers. When a utility purchases water from a bulk supplier, it is less vertically integrated. *Economies of vertical integration* are also labeled *economies of sequence*.
- *Isoquants:* when characterizing production functions graphically, the isoquants represents combinations of inputs that yield the same output.
- *Elasticity of Input Substitution* captures the extent to which inputs can be substituted for one another.

Isoquants can be used to illustrate how inputs might substitute for one another. Let us take the production function used earlier, $Q = N(10L - 0.2L^2)$ to graph the different combinations of Labor and Network size that can be used to produce a given level of output (Figure 2).

The isoquant labeled Q = 30 shows that 30 units of output can be produced by using the combinations of N and L shown by points **a**, **b**, **c**, and **d** (or, assuming that inputs can be continuous, on other points shown in the Q = 30 isoquant). Similarly, 120 units of output can be produced by using 6 units of L and 2.27 units of N (as shown by point **h**), 3 units of L and 4.25 units of N (as shown by point **f**), or other point on the isoquant corresponding to Q = 120.

Note that the production function that we have been using allows for the firm to substitute one input for the other (the curvature of the isoquant shows us the extent to which the inputs can be substituted for each other). For example if the utility were producing 120 units of output at point **e** using 6 units of labor and 2.27 units of network and the price of labor dramatically increased but the cost of capital (associated with the network) fell, managers would be able to produce the same amount of output using less labor: they could instead use 3 units of

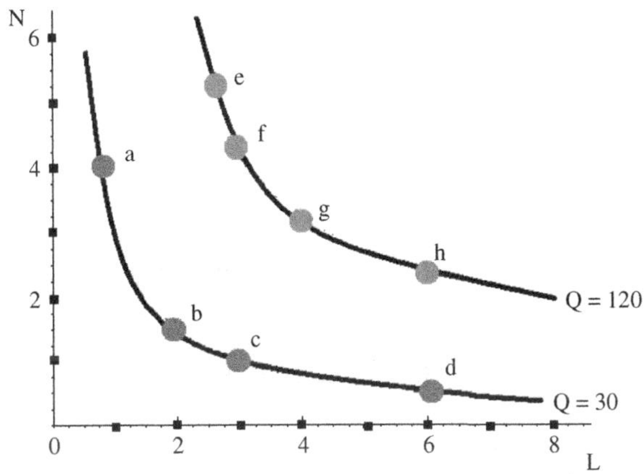

Figure 2. Production Isoquants

labor and 4.25 units of network. This movement along the isoquant could not be accomplished instantaneously. In the short run, management can only vary labor. In a more complete production model, the rate of use of energy, chemicals and other variable inputs can be altered fairly easily, but capacity expansion (illustrated her by N, network size) will require planning, bidding, installation and other tasks that require more time.

Of course, this example identifies a problem with the specific production technology being used here to illustrate economic concepts. Firms may *not* be able to substitute their inputs for one another, even over the long term. Delivering water to 4000 customers at a particular set of locations requires a specific network of pipes and pumps. The manager cannot use more labor (or other variable inputs) and reduce the amount of capital (network). Such processes are said to have technologies with "fixed factor proportions" (zero elasticity of substitution). Many stages of a water utility's production process have the characteristic. The case of bottled water illustrates this principle: the inputs that go into producing bottled water, plastic and water must be used in a fixed proportion. By using 1 bottle and 0.5 liters of water, a firm can produce one bottle of water. If the firm has 1 bottle (N) and 1 liter of water (L) it can still just produce one bottle of water (similarly, 2 bottles and 0.5 liters allow for 1 bottle to be produced). This can be represented graphically by the isoquants shown in Figure 3. Note that the only points that are technically efficient are the ones shown by the dots (the dashed lines show inefficient production points).

On the other hand, billing could be automated (using many computers—capital—and only a few workers) or it could be labor intensive. Furthermore,

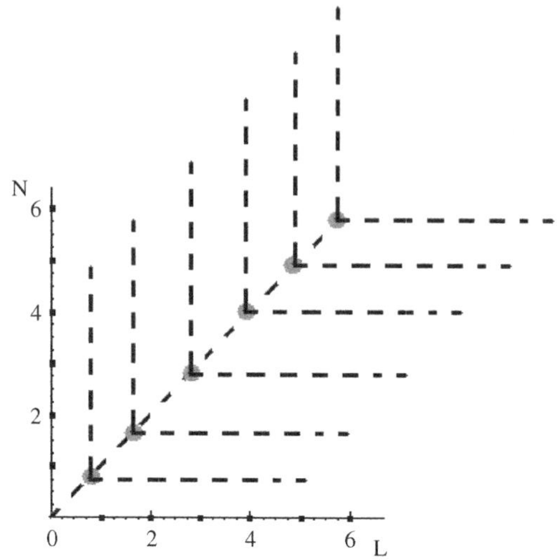

Figure 3. Production Technology with No Input Substitution Possibilities

once the utility has a network of a specific size, if customers demand more water (as family sizes increase or incomes grow), the utility can expand output by using more variable inputs (e.g. labor, energy and bulk water). So the short run production technology does not exactly match that depicted in Figure 3. Thus, the overall production function estimated for a utility will be a mixture of many processes. The results will approximate reality, giving some insights into scale economies and substitution possibilities as revealed by the sample observations. Estimated coefficients and the associated efficiency scores do *not* help managers identify specific steps in the production process that might be inefficient, but efficiency scores based on production functions can provide useful information for ranking firms in the aggregate.

3.2 Statistical Estimate of a Production Function

Drawing upon the previous example where L ranged from 1 to 5 and N (network size) was either 1 or 2, let N vary from 1 to 4, so we have 20 possible output levels associated with the combinations of inputs. We will create a dataset from this specific function:

$$Q = N(10L - 0.2L^2)$$

Thus, we could find an approximation of this functional form by using the data to estimate another function, one that is often utilized by researchers, the Cobb-Douglas production function. Note that the resulting coefficients for this Cobb-Douglas model will represent the "best fit" of the model for data over this particular range of inputs and outputs. If we had observed (selected) data associated with much greater output levels (based on the "true" production recipe, the estimated coefficients for this approximation of the "true" model would be very different. For now, we consider the twenty output levels (for L = 1 to 5 and N = 1 to 4), and show how coefficients can be derived and interpreted.

The Cobb-Douglas production function is a production function of the form:

$$Q = AN^{\alpha}L^{\beta}$$

where
Q is output,
A is a productivity (or technology) index,
N is network size (capital), and
L is labor.

Note that the form is multiplicative, with the two inputs (N and L) having exponents. The A term has an intuitive interpretation as an efficiency index: holding all inputs constant, if A is 10% larger, then output also is 10% greater. If the analyst has data for a number of firms over several years, the efficiency might be shown to change over time—reflecting technological change (the capacity to obtain more output from a fixed amount of inputs). For our purposes, we will estimate A as a constant using the twenty "observations". The exponent terms also have convenient interpretations. Holding everything else constant, we would compute the impact of a 1% increase in L as β. Similarly a 1% increase in N would increase Q by α. If both inputs were increased by 1%, total output would increase by $\alpha + \beta$. So the sum of the exponents for a Cobb-Douglas production function provides useful information: there are increasing returns to scale if $\alpha + \beta > 1$.

To obtain estimates of the three coefficients, A, α, and β, the analyst can transform the function by taking the natural logarithms of both sides of the equation relating Q to the inputs. The underlying relationship is unchanged. Allowing for an error term (u) in the transformed relationship, this expression can be rewritten as a linear equation to be estimated[1]:

$$\ln Q = \ln A + \alpha \ln N + \beta \ln L + u \text{ (where } \ln A \text{ is the intercept).}$$

[1] Using the following rules of logarithms:
(a) $\ln(x) + \ln(y) = \ln(x*y)$
(b) $\ln(x)^a = a \ln(x)$

The parameters lnA (intercept), α, and β can, thus, be estimated by linear regression, where the steps are shown below:

(1) Taking the natural logarithms of Q, N, and L, yields in the following data (presented in Table 1):

Table 1 Output and Input Data

Q	L	N	Ln(Q)	Ln(L)	Ln(N)
9.8	1	1	2.282	0.000	0.000
19.2	2	1	2.955	0.693	0.000
28.2	3	1	3.339	1.099	0.000
36.8	4	1	3.605	1.386	0.000
45.0	5	1	3.807	1.609	0.000
19.6	1	2	2.976	0.000	0.693
38.4	2	2	3.648	0.693	0.693
56.4	3	2	4.032	1.099	0.693
73.6	4	2	4.299	1.386	0.693
90.0	5	2	4.500	1.609	0.693
29.4	1	3	3.381	0.000	1.099
57.6	2	3	4.054	0.693	1.099
84.6	3	3	4.438	1.099	1.099
110.4	4	3	4.704	1.386	1.099
135.0	5	3	4.905	1.609	1.099
39.2	1	4	3.669	0.000	1.386
76.8	2	4	4.341	0.693	1.386
112.8	3	4	4.726	1.099	1.386
147.2	4	4	4.992	1.386	1.386
180.0	5	4	5.193	1.609	1.386

(2) Here, ln Q is thedependent variable, and lnN and lnL the explanatory (independent) variables. The intercept term is lnA. We can now estimate the model,

$$\ln Q = \ln A + \alpha \ln N + \beta \ln L + u,$$

by regressing lnQ on lnN and lnL, and, therefore, find the intercept and coefficients (i.e. lnA, α and β).
The Ordinary Least Squares estimate of the transformed Cobb-Douglas model yields the following:

$$\ln Q = 2.289 + 1 \ln N + 0.949 \ln L$$

where
Constant term = lnA = 2.29; α = 1; and β = 0.95.

(3) From the rules of logarithms we know that $A = e^{\ln A}$, which implies that:

$$A = e^{2.289473} \approx 9.87$$

(4) Substitute these back into the original function ($Q = AN^{\alpha}L^{\beta}$) to obtain estimated Cobb-Douglas relationship:

$$Q = (9.87)(N)^{1}(L)^{0.949}$$

Note that $\alpha + \beta = 1.949 > 1$, which implies increasing returns to scale *over this particular range of inputs and outputs*: if we increase both N and L by 10%, Q increases by more than 10%, in this case almost 19.5%. The "fit" of the estimated model is very good. For example, we know from the table that when L = 4 and N = 3, Q = 110.40. If we insert L = 4 and N = 3 into the Cobb Douglas approximation of the production function that was found in step 4, we obtain $Q = (9.870)(3)^{1}(4)^{0.949} = 110.355$ which is very close to the Q that was "observed" (110.40, found in the original dataset). Note that had we estimated the relationship over a different output range, the returns to scale would have been very different, which underscores the importance of testing for variable returns to scale, to determine the extent of scale economies over different ranges of output. Other functional forms would allow the analyst to test for non-constant returns to scale.

3.3 Cost Concepts

Outlays (or utility expenditures) capture some of the costs associated with water delivery. Economists utilize terms like total cost, average cost, and marginal cost to characterize different aspects of cost. Economies of scale, scope, sequence, and density refer to long-term adaptations of output levels and output mixes to the underlying production technologies and associated demand conditions. Clearly, the process is evolutionary in nature, as managers discover how the operating characteristics of the organization limit (or enhance) various activities. Engineering cost functions can put boundaries on costs, but operations in a given time period (and the impact of the learning curve over a number of years) need to be understood if cost data are to be meaningful. Accounting information provides useful historical data, but these numbers do not necessarily capture economic reality.

- *Total cost* (TC) depends on total output (holding call input prices and the production technology constant)
- *Average cost* (AC = TC/Q) is also called unit cost: cost per unit of output

- *Falling (or declining) Average Cost* can be understood in two very different ways. For network services that require substantial fixed investments, short run average cost falls so long as there is excess capacity. In the short run, average cost falls since the cost of producing additional output is relatively low. Non-specialists sometimes confuse such "spreading fixed costs over a larger output" with scale economies (see below).
- *Scale economies* is a long run concept. Here, declining average cost stems from the underlying production technology. Capacity is allowed to expand; economies of scale result when the firm obtains larger percentage increases in output than the percentage increase in inputs. For example, if an increase in total output of fifteen percent causes total costs to rise by ten percent: average cost falls.
- *Marginal cost* (MC = dTC/dQ) is the opportunity cost of producing one more unit of output. The time frame affects the calculation of marginal cost, since capacity cannot be adjusted in the short run. Note, if water is "free" yet the water table or surface water source is being drawn down then the economic definition of MC would include the future consequences of that action. For example, if the utility will have to go a much more expensive source in the future (like desalinization), ignoring the opportunity cost of consuming water today leads to very high future costs.
- *Incremental cost* can be used to approximate MC if a TC function is not available. It is the change in total cost caused by a discrete increase in output divided by that increment of output.
- *Short run marginal cost* is calculated holding production capacity fixed.
- *Minimum efficient scale* gives us an idea of how many suppliers could "fit" into the market. If demand is low relative to minimum efficient scale, we would conclude that the supplier is a *natural monopoly*. Most empirical studies of water systems find moderate economies of scale relative to the typical market size.
- *Break-even analysis* can help decision-makers understand the level of sales required if the firm is to cover its costs (including a normal return on investment). For a pre-determined price, the volume of break-even output will be higher if total costs are higher. This decision-framework is especially helpful if *demand is uncertain*, and managers need to determine the sensitivity of cash flows to different cost structures.

An analyst can combine production information (shown in the Figures in the previous section) and input price information by using both isoquants and isocosts. Isocosts show different combinations of labor and network size that have the same cost: input prices determine the placement of an isocost line. Note that Total Cost in the long run is given by

$$TC = wL + rN$$

Where w is the price of labor and r the "price" of a "unit" of network. The latter can be interpreted as a rental price or an annualized cost reflecting the cost of initially acquiring and installing the pipe multiplied by the effective cost of capital (say 12%) to obtain an annual cost.[2] As an example, in Figure 4 below, the line labeled c = 3 shows all the combinations of labor and network that cost \$3, assuming that w = 1 and r = 3, where w is the cost of labor (wages) and r is the cost of "renting" each "unit" of network (as with a mile of pipe). Three isocost curves are depicted below:

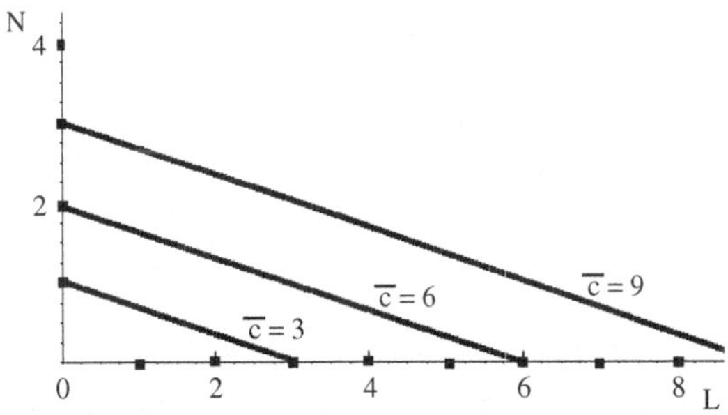

Figure 4. Isocost Curves

We can combine the isocost graph with the isoquant graph to find the combination of inputs that minimizes costs for a given level of output. Given our original function, $Q = N(10L - 0.2L^2)$, an output of 30 units, and assuming w = 1 and r = 3, the ideal point of production is c with a cost of \$6, since any other point on the same isoquant, such as point **d**, would be more expensive (producing 30 unit of output at point **d** costs \$7.70). Figure 5 shows that the cost of producing Q = 30 at either points **a** or **d** involves a higher total cost: these input mixes would be on isocost lines passing through (not tangent to) those points.

[2] For now, we take a very simplistic view: r is taken to be a rental price. In fact, the capital component for cost in water utilities involves some very complicated measurement issues including the economic lifetime of the asset.

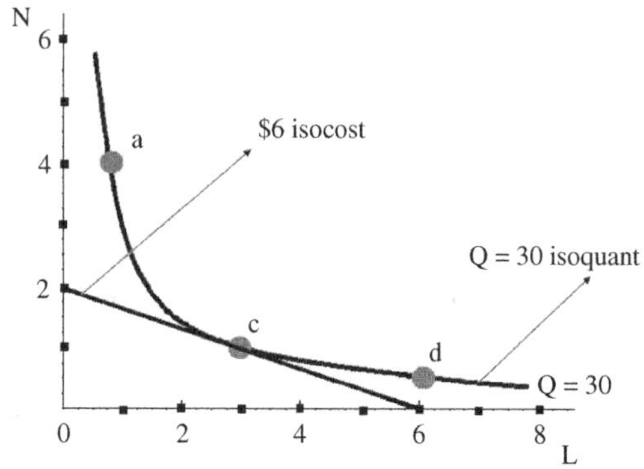

Figure 5. Efficient Input Mix

The calculations supporting this observation are shown below.

Point	Q	N	L	r	w	wL	rN	TC = wL+rN
c	**30**	**1.06**	**3**	**3**	**1**	**3**	**3.18**	**6.18**
a	30	4.06	0.75	3	1	0.75	12.18	12.93
d	30	0.56	6	3	1	6	1.70	7.70

A line connecting all of the tangency points for the isoquants and isocost lines is called the *expansion path*, shown in Figure 6. The expansion path allows us to derive the long run total cost curve, since the path contains the cost-minimizing combinations of labor and network size that are required for each level of output. Note, the expansion path (and total cost for each Q) will depend on the input prices. If each input price doubled, the path (tangencies) would be unchanged, although total cost would double for each output level. If the relative prices changed, as with a reduction in r and an increase in w, then the expansion path would shift to the Northwest—along a more capital-intensive (network-intensive) path. The shift would reflect input substitution.

Thus, production functions focus on how inputs affect output (or outputs). Cost functions consider how output levels and input prices affect total cost. Both can be used to evaluate relative performance of water utilities.

3.4 Efficiency Scores Reflecting Outputs and Inputs

This section provides a more extensive discussion of outputs and inputs. We will use "capital" as the indicator of network size, and continue to use labor as the

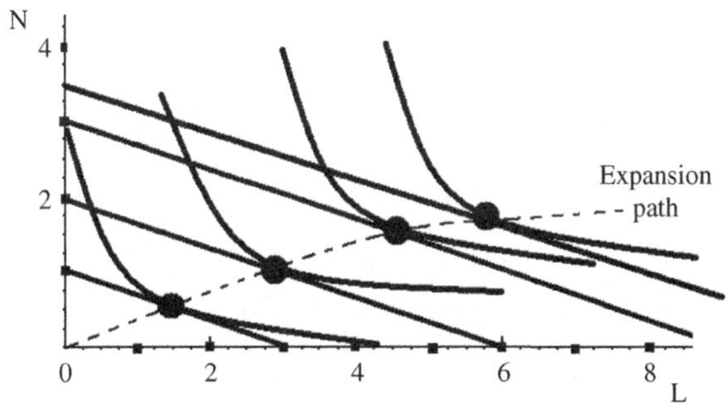

Figure 6. Optimal Input Combinations

only variable input. The example is highly stylized, but illustrates how the concepts enable the calculation of efficiency "scores". Imagine a hypothetical dataset as follows, where we have obtained data on the input and output levels of eight different water utilities that deliver water using a standard technology:

Utility	Capital	Labor	Water Delivered
U1	40	30	10
U2	70	30	10
U3	80	10	10
U4	40	20	10
U5	20	40	10
U6	50	20	10
U7	60	40	10
U8	55	25	10

Having the same total output (water delivered) simplifies transferring the data onto a graph (Figure 7). If the output levels had differed, we would "normalize" the graph, dividing the inputs by the output and showing the unit output points on the graph. Graphing these data points we can analyze different concepts of efficiency and intuitively explain what they mean.

Figure 7 shows how these eight firms will be compared against each other in terms of their technical efficiency levels. This means that each firm is going to be analyzed regarding how much output it is producing relative to the amount of inputs it is using, assuming that the same technology available to all firms. The last condition implies that hydrology and other conditions are the same or have minimal impact on production. From this perspective, if a firm is producing the same output as another firm and it is utilizing fewer inputs in the process, the

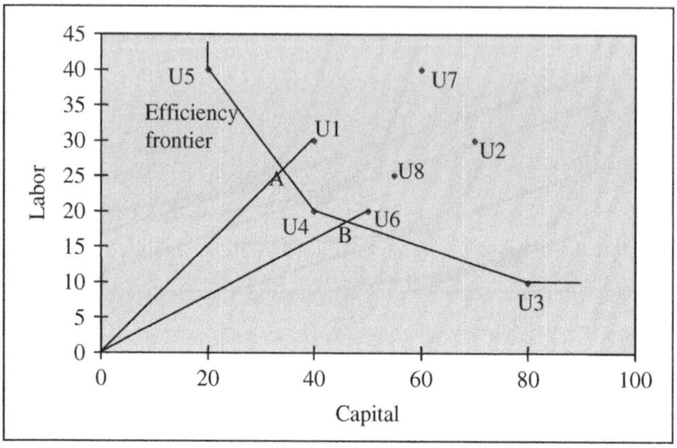

Figure 7. Efficiency Frontier and Technical Efficiency

former can be labeled as technically more efficient than the latter. This framework allows us to measure the technical inefficiency of one firm with respect to others.

The figure shows eight firms that produce the same amount of output by combining different amounts of labor and capital, which are the only two necessary inputs this production technology requires. Since the firms are utilizing different combinations of both inputs, it seems difficult to compare how well they are performing in terms of technical efficiency. This task would be much easier if there were only one input utilized in the production process. However, an "Efficiency Frontier" can be constructed with the input data: the frontier relates the minimum amount of inputs required to produce a given output level. With that tool in hand, firms can be compared against each other even when they use a different combination of labor and capital to produce the same amount of output.

3.4.1 Efficiency frontier

When this Efficiency Frontier is constructed by mathematical programming methods (like the one used by DEA), a line (frontier) connecting the points that are "closer" to the origin is found such that no point lies between the line and the origin. This line generally has multiple kinks, and it should also slope downward, showing the substitution possibilities between the two inputs. That is, the same output level can be produced with less labor ONLY if more capital is used. This technique allows the analyst to compare all the firms, by observing how far they are from the "frontier." For example the efficiency score of U7 will be lower than U1, since it is further from the frontier. All the firms in the sample can be given a technical efficiency score even when each firm is unique in terms of using different amounts of capital—or network size—for example. In the simpler case, if each firm had the same amount of capital, then we could compare the amount of labor used, concluding that the firm using more labor is technically less efficient than another.

In the set of utilities depicted in Figure 7, a line is constructed that connects firms 3, 4, and 5. These three firms determine the Efficiency Frontier in this case. By joining these 3 points, all the firms in the sample are categorized into two groups: on the frontier or away from it. The three firms used to construct it are *on the frontier* and no technical (engineering) inefficiency can be assigned to them. The frontier is equivalent to a non-linear isoquant: output is the same at each point on the frontier. The remaining firms outside the frontier are labeled as *technically inefficient.* They could reduce the amount of inputs used and still produce the same amount of output as the firms on the frontier.

For example, if firm 1 moved to point A (a combination of labor and capital corresponding to moving along a ray to the origin), it would reduce both the amount of labor and the amount of capital utilized, and still produce the same level of output. Thus, firm 1's efficiency score is relative to its "efficient twin" (a virtual firm) that is producing on the frontier (at point A). The same process can be applied to firm 6, which could move to point B by using a lower capital and lower labor combination of inputs to still produce its actual output level.

3.4.2 Calculating technical efficiency

Notice that in the conclusion about where a firm should move to be technically efficient, we identify a "virtual" firm on the frontier where no firm operates. We assume that firm 1 "could" produce at point A and that firm 6 could have used the labor capital combination depicted by point B, even though no firms are actually located at these points. This frontier is constructed by assuming that if a firm locates at A or B, it will be able to produce the same amount of output that firms 3, 4, and 5 are producing.

We are now in a position to formalize how the technical efficiency score is calculated, using firm 1 to illustrate the calculation. Since that firm could move to point A and still produce what it is now producing, the distance from the original point (U1) to A is the "reduction" in inputs that it should achieve to locate on the frontier. That distance reflects how "far" that firm is from the frontier. It is natural to compare that distance with the distance between the origin and the point where the firm is now operating. Technical efficiency is defined found by dividing optimal distance by the actual difference. For firm 1, the distance from A to U1 reflects the size of the "reduction" in inputs if the firm moves to the frontier. The technical efficiency score is defined as the ratio of the segments OA/OU1 < 1 .Similarly, the technical efficiency score for firm 6 is denoted by OB/OU6 < 1. Note that if we perform that calculation for firms 3, 4, and 5, we obtain an efficiency score of 1.

3.4.3 Calculating allocative efficiency

The same eight firms that were analyzed in terms of their technical efficiency levels can also be compared in terms of a different efficiency concept: *allocative efficiency* (also called production efficiency or cost minimization)[3]. This new efficiency concept focuses on how well the firm accounts for input prices in its production process. The technical efficiency concept only analyzed how much output the firm was able to produce from a given basket of inputs. It only focused on the technological or engineering opportunities, ignoring input prices. This additional efficiency concept is useful for evaluating the combination of inputs used, even when the firm may be technically efficient. That is, even when the firm may be satisfied with its technical efficiency score (because she cannot reduce inputs and still produce its actual output level), the allocative efficiency concept enables managers (and external monitors) to evaluate whether costs are being minimized for producing a given level of output. By incorporating input prices into the analysis, the framework identifies a different input combination that keeps output constant but reduces the total expenditures by the firm. For example, we indicated that firm 1 could move to point A and achieve technical efficiency. Such a change in its input use brings the firm to the technical efficiency (TE) frontier: earning a TE score of 1. However, when the allocative efficiency concept is applied, it shows that the firm could do even better. Figure 8 allows us to distinguish between the two types of efficiency.

To distinguish between the two types of efficiency, we need to introduce the *isocost curve* into the analysis (seen in Figure 8). Recall that this straight, downward sloping line reflects all the possible combinations of capital and labor that can be achieved by spending the same amount of money: iso or

[3] To be completely rigorous, note that full allocative efficiency also requires that price be set to marginal cost, so the price signal to consumers reflects the opportunity cost of supplying additional output. Of course, since the short run marginal cost of additional water is very low (and long run bulk water costs could be very high), the correct price signal is involves balancing short and long term considerations. We do not examine the area of rate design since that would take us far afield—into hook-up charges, monthly fees, seasonal rates, and block structures.

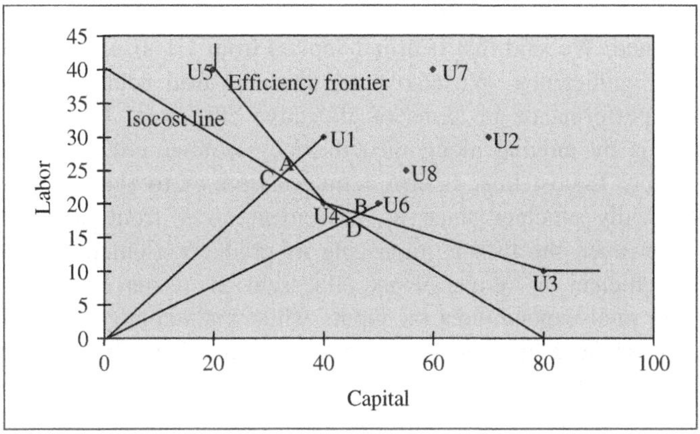

Figure 8. Allocative Efficiency and Technical Efficiency

equal cost. The slope of that line reflects the relative prices of inputs. By fixing total expenditures at some level (say, $80), we can easily construct the line by joining two points. One point is on the horizontal axis, found by dividing the total expenditures of $80 by the price of capital (assumed to be $1 per unit of capital in our graph). The second point will be on the vertical axis, and it is found by dividing the total expenditures by the price of labor ($2 in our graph). Clearly, those two points identify the maximum amounts of each input that can be purchased if the firm decides to spend all the money ($80) on one input. Hence, the isocost line connecting these two points tells us how we can move from a point where the firm buys only labor to a point where the firm buys only capital, while keeping total expenditures constant ($80). Points C, D and U4, for example, show us three different combinations of labor and capital that can be purchased by using the same expenditure level. These points lie on the isocost line, as well as (80, 0) and (0, 40) as has been already mentioned.

Changes of input prices will change the slope of the isocost line. For example, if capital was suddenly cheaper the firm could get more than 80 units of it without changing its total expenditures. On the other hand, a lower price of labor would allow the firm to get more than 40 units of labor when spending all the money on that input. Changes in total expenditures, on the other hand, do not change the slope of the line but they shift its position on the graph. A higher expenditure level allows the firm to get larger quantities of both inputs, hence moving the isocost line away from the origin.

Once the isocost line is introduced, the concept of allocative efficiency is easily explained. We said that if firm 1 moved from U1 to A, it eliminated all its technical inefficiency. We also noted that the firm could go further and improve its performance in terms of allocative efficiency. The way the firm can do this is by moving along the efficiency frontier until she gets to the lowest possible isocost line. If firm 1 moves from A to U4, the firm is still being technically efficient since no movement away from the frontier has occurred. However, the firm is improving its production efficiency, since it is now more efficient in terms of its allocation of resources, because it is reducing the total expenditures on inputs while maintaining the same output level. Of course, such movements are long term, since they involve changes in capital. That may be a strong assumption for a utility network built when input prices were different. Nevertheless, the concept lets analysts further evaluate how utilities with perfect foresight regarding input prices might diverge from having an appropriate mix of inputs—at least in principle.

Thus, allocative efficiency measured by is comparing the cost actually incurred with the (presumably lower) cost that could be attained by utilizing a different input combination, while keeping output unchanged—somehat along the lines of the engineering "hypothetical firm" approach to performance comparisons. Here, the ideal cost is dictated by the isocost line. For example, a measure of allocative inefficiency for firm 1 is found when comparing the potential cost savings (segment CA) with respect to the actual cost (segment OA). The same reasoning applies if we want to find the allocative efficiency score for the same firm: the score is given by the ratio of the segment OC (lowest possible cost for that level of output) and the segment OA (actual expenditure). Then, the ratio OC/OA < 1 is the allocative efficiency score for firm 1. Similarly, the ratio OD/OB < 1 is the allocative efficiency measure for firm 6. Examining the performance of firm 4, one observes that there is no possible cost reduction, holding while output constant. There is no movement along the efficiency frontier that would allow the firm to spend less on inputs and at the same time produce the former level of output. Then, the conclusion is that firm 4 is allocatively efficient: its ratio is OU4/OU4 = 1.

Economists like to be precise, although the refinement may not be crucial for a regulator attempting to rank firms—especially when input price data are difficult to obtain. Nevertheless, separating performance into technical and allocative (production) efficiency allows the introduction of a broader concept: total efficiency. We can keep utilizing firm 1 as an example, as follows. Firm 1 is producing at point U1. At that particular point we found a level of inefficiency given by the segment AU1 (potential input usage savings), and a level of allocative inefficiency given by the segment CA (potential expenditures savings). Intuitively, the combination of both inefficiencies accounts for the total inefficiency that firm 1 is experiencing. Formalizing this concept into efficiency scores, we have technical efficiency TE = OA/OU1 and allocative efficiency AE = OC/OA. The product of the two gives the total economic efficiency EE = OC/OU1 (after the cancelling OA in the numerator of one term and the denominator of the other). For firm 6 for example, we have TE = OB/OU6 and AE = OD/OB, both composing a score of EE = OD/OU6. The same can be done for all the firms in the sample.

Note that firm 4, which we already found to be technically efficient has no potential cost savings, since it is on the isocost line. Then, firm 4 can be labeled as the only economically efficient firm in the sample. It has a score of 1 in both the TE score, as well as the AE score. Thus, being economically efficient necessarily implies being technically efficient. But the reverse is not true: even if a utility technically efficient, it could still be economically inefficient because of a score less than 1 regarding allocative efficiency. Such refinements are more in the domain of academic studies, but the ideas identify directions for future work.

3.5 Outputs and Costs

It is useful to further develop efficiency notions in the context of costs. After considering scale efficiency, this section describes statistical techniques for estimating a linear cost relationship. It is possible that the data might support a more complicated functional relationship. However, alternative cost model specifications will not be explored here.

3.5.1 Scale efficiency: the concept

For the final efficiency concept that will be introduced, *scale efficiency*, we introduce a new hypothetical dataset showing several firms producing different output levels at different operating costs:

Utility	Operating Cost	Water Delivered
A	20	10
B	40	20
C	30	30
D	50	40
E	55	30
F	60	50

Figure 9 shows this situation graphically, with output measured on the vertical axis and costs measured along the horizontal axis:[4]

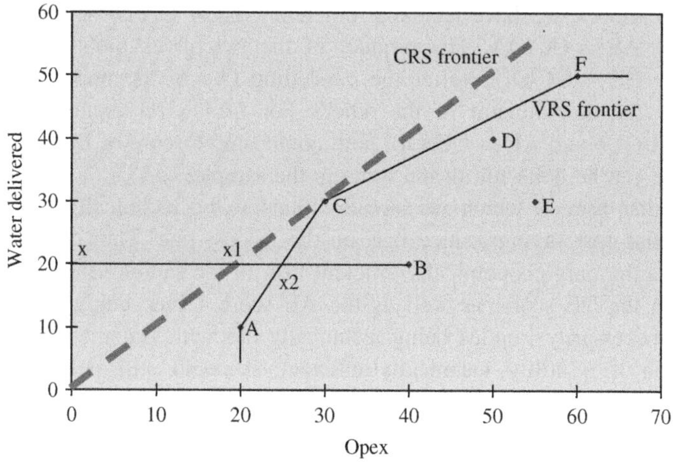

Figure 9. Scale Efficiency

The scale efficiency concept is important in the water industry, since firms are usually considered to face economies of scale, meaning that they could double the amount of output produced without doubling the cost of producing it. In such situations, it would be useful to measure (and control for scale) when studying efficiency across water utilities. If scale is beyond the control of managers, then a small firm with high unit costs should not be penalized. The proper comparison would be with a comparably sized firm.

Analysts seek to find a frontier in order to compare the performance of each firm with the ideal one. The mathematical programming techniques used by DEA would construct such a frontier by finding a line that leaves all the utilities either on the frontier or below it. Then, a firm that ends up on the frontier is said to have no scale inefficiency, while a firm below the frontier is suffering from scale inefficiency. Of course, those higher costs are presumably beyond the control of the manager, though the presence of such inefficiencies could point to industry consolidation to gain economies in some aspects of the production process.

[4] Generally, cost would be on the vertical axis, with output on the horizontal axis, but this format illustrates how obtaining more output from a given level of outlays affects the frontier.

An important distinction has to be made regarding the way the frontier is constructed. If the analyst considers that the industry is characterized by constant returns to scale (CRS) (i.e., doubling output would imply doubling costs), then the frontier is constructed as a straight line joining the origin with the firm depicting the highest ratio water/cost. In Figure 9 that line is the dashed line passing through point C. But if the analyst considers that the industry is characterized by variable returns to scale (VRS), a different frontier has to be constructed (the kinked one in the graph). This frontier still has many firms off the frontier, but some firms look better now since they are now on the frontier. Which of the two approaches is utilized is a decision residing on the analyst, who will judge (and empirically test) which framework is more appropriate. Of course, the scale efficiency score obtained for each particular firm will differ under the two approaches (except for firm C, the one picked to construct the CRS frontier).

3.5.2 Measuring scale efficiency

Measuring the scale efficiency score is similar to making computations described for production functions. Intuitively, it is just a ratio of two distances on the graph: what it "should be" over what it is. The score will be then less than 1 in case the firm has something to improve, but it will be equal to 1 when the firm is producing efficiently. Utilizing firm B as an example, a scale efficiency analysis under the CRS assumption indicates an efficiency score of $xx1/xB < 1$. This scale efficiency score less than 1 says that the firm is not operating efficiency (according to the CRS assumption). According to the output level of 20 units produced, the firm should have a cost of 20 rather than the cost of 40 actually observed (and reported in the table above). On the other hand, given a VRS assumption, firm B faces a scale efficiency score of $xx2/xB < 1$. This score is still less than one still (indicating inefficiency for this utility): since costs are larger than what they should be for this output level. However, the VRS assumption is more lenient, and suggests a potential cost reduction to only 25 units (relative to the actual cost of 40). Thus, one can see why the "returns to scale" assumption is so important. Under the CRS assumption, only firm C is labeled as being scale efficient, while under the VRS assumption, firms A and F are also considered as scale efficient. The potential outputs under VRS are lower for any given level of costs. This refinement illustrates the importance of model specification. Here, constant returns to scale imply very different levels of inefficiency relative to variable returns to scale—here decreasing returns to variable inputs, presumably because fixed inputs are being held constant.

3.6 Statistical Estimates of a Linear Cost Function

The table below shows hypothetical data on cost and water delivered from a sample of 8 utilities.

Utilities	Water Delivered	Cost
A	3	18
B	3.5	20
C	4	10
D	5	27
E	6	36
F	7	53
G	8	33
H	9	62

Figure 10 is constructed based these data, with Cost appearing as the dependent variable on the vertical axis, and water delivered (the independent variable) on the horizontal axis. The Figure shows how three different cost frontiers would look, depending on which statistical approach (model) is taken to derive the relationship: Ordinary Least Squares (OLS), Corrected Ordinary Least Squares (COLS) and Stochastic Frontier Analysis (SFA).

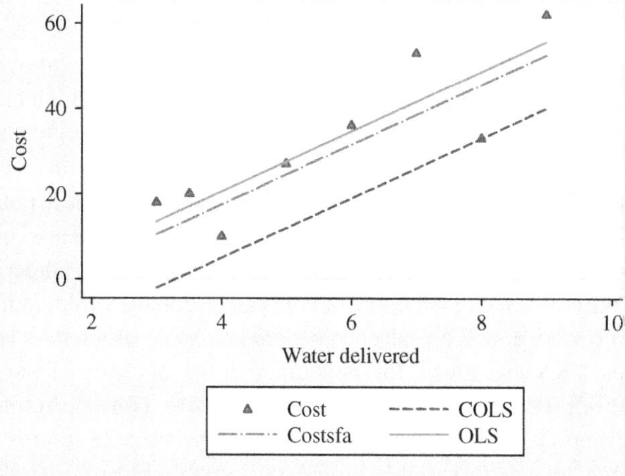

Figure 10. Comparison of OLS, COLS, and SFA Linear Relationship

As can be seen from Figure 10, the three methodologies utilize the same information from the firms to construct different regression lines. Assumptions regarding the error terms (and their distributions) lead to different estimates of the relationship between water delivered and cost. The OLS frontier is the top one, while the COLS frontier is the bottom one. The SFA cost frontier lies between the two other regression lines.

One can see that the OLS line is the least severe of the three when evaluating efficiency, while the COLS line is the one that punishes firms the most in terms of calculated efficiency scores, where the "score" would be the predicted cost divided by the actual cost. Notice that for any firm in the graph, the vertical difference down to the cost line (cost "standard") chosen depicts the potential cost savings that the firm could achieve.[5] This potential cost savings is labeled as the firm's inefficiency, depending on the performance of other firms and the methodology chosen to rank performance. Then, any firm will be considered more efficient when using OLS than when using SFA, and also, any firm will be considered more efficient if SFA is used rather than COLS.

A complete and detailed explanation about why each line is placed in a different position on the graph is beyond the scope of this volume. However, it is helpful to provide some intuition regarding the methodologies. OLS is a technique that focuses on the "average:" observations can be above or below the predicted "line". Deviations from predicted cost can reflect performance that is above or below average. The COLS approach is almost the same as OLS, with the difference that it picks the best of the firms (from OLS) and assumes that *this firm* is 100% efficient. Then, the cost line goes through *that* observation and all the remaining firms in the sample are considered inefficient relative to that firm. So the two approaches use different baseline "standards." The calculated efficiency scores will differ, although the rankings of the firms would be the same!.

SFA partially alleviates the extreme assumption of COLS, providing a measure of inefficiency between OLS and COLS. However, the result comes at the cost of additional assumptions (regarding error terms) and the use of a more complicated statistical package. SFA still assumes that part of the deviation from minimum cost is attributed to inefficiency (as COLS), but it also recognizes that part of that deviation can be considered to be a random disturbance (i.e., statistical noise or measurement error). Thus, SFA takes away part of the difference between observed and estimated (ideal) cost, ending up with smaller estimates of inefficiency. For regulators and managers just beginning the

[5] In case the observation is below the line, as with roughly half the observations for OLS, the firm is performing "better than expected: its cost is less than predicted".

exploration of models to characterize efficiency, it is probably best to utilize simple models until (and unless) more complex formulations do a better job of explaining relative utility performance.

3.7 Specification of a Nonlinear Relationship

Consider the same utilities used in the previous example but with two changes in reported costs: Utility F has total cost of 40 and Utility H has total cost of 45.

Utilities	Water Delivered	Cost
A	3	18
B	3.5	20
C	4	10
D	5	27
E	6	36
F	7	**40**
G	8	33
H	9	**45**

Figure 11 shows the new observations. Now, it appears that the linear relationship that would have been estimated using OLS, COLS, or SFA is no longer present. The relationship between production volume (water delivered) and total cost is now nonlinear.

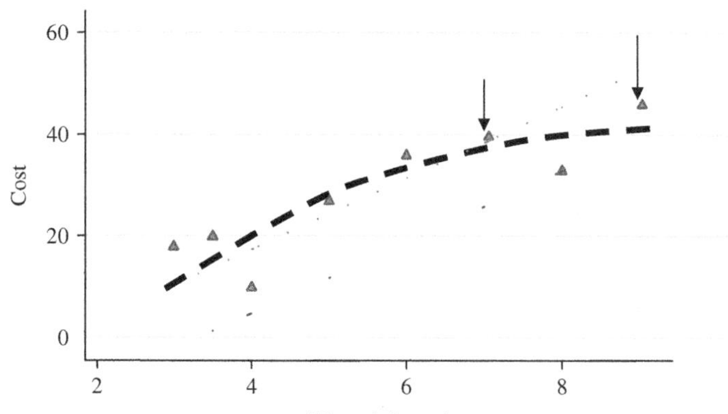

Figure 11. Possible Nonlinear Cost Function (due to changes in two observations)

From Figure 11, we can see that total cost still rises with increased volumes of water, but cost does not rise proportionately. Such a relationship would indicate economies of scale if the observations refer to utilities of different production capacities. Suffice it to note, that model specification should take potential nonlinearities into account.

Chapter 4 Strengths and Limitations of Different Methodologies: Technical Considerations

The preceding section provided the intuition behind several statistical approaches to modeling production and costs. The objective of this section is to examine in greater detail the strengths and limitations of different methodologies and describe ways to determine the robustness of performance rankings. The Benchmarking Checklist summarized this material, but it is useful to have a more comprehensive basis for evaluating alternative methodologies.

4.1 Criteria for Selecting Performance Measures

Before considering the different methodologies that have been introduced through examples and graphs, we need to review the principal *criteria for choosing* a set of performance measures: they should be understandable, comprehensive, useable, and timely.

- ***Understandable:*** The measures selected should facilitate the development and integration of essential information on the company's overall performance. It is not very useful to have dozens of measures if there is no logical framework for integrating them into a measure of overall performance.
- ***Comprehensive:*** A large number of productivity measures may provide a comprehensive picture of a firm's behavior but it is typically difficult to determine the accuracy of rankings or scores. Thus, the performance score is not readily comprehended and may create confusion if some indicators move in opposite directions. Comprehensive summary measures are needed, with explicit weights given to different components.
- ***Useable:*** The intended audience must understand the indicators and how they are affected by external developments (beyond the WSS utility's

control) and by internal actions. Basically, the information must be decision-relevant—useful for managers, regulators, and policy-makers.

- ***Timely*:** Poor performance in the past may have been already corrected, so having current information is necessary to assess actual performance. Of course, information on trends puts current data into perspective, enabling decision-makers to implement reforms, design incentives, and evaluate the impacts initiatives.

No single approach provides a comprehensive index of overall performance. However, one could argue that production and cost function studies have been underutilized by regulators and managers.

4.2 Specific Core Indicators (Partial Metric Methods)

Concept: The partial metric method consists of calculating different measures of the financial, operating, commercial, and quality characteristics of a firm. Performance indicators can be based on cross-sectional comparisons with competitors or comparable firms, while past performance can provide information on improvements over time. In general, these indicators incorporate relationships between two simple measures, yielding indices of productivity, leakage and theft, human capital development, or financial conditions, among others. When presenting the benchmarking results it is important to tailor the set of indicators to the purpose or to the intended audience. For instance, managers may view some of the indicators as not under their control. Hence, it is helpful to supplement the main indicators with summary measures to reinforce the effect of possible improvement actions. Within this group, we can identify Specific Core Indices, and Overall Performance Indicators (OPI).

A single index of utility performance has the same problems of any indicator: it will be neither comprehensive nor fully diagnostic. As was noted earlier, physician can collect information on a patient's temperature, pulse, height and weight. The indicators point to potential or existing health problems. A fever is a short term problem that can be addressed with specific medications; weight is a longer-term health issue with implications for potential heart attacks and other problems—diet and exercise programs might be prescribed. However, a set of blood tests will provide more detailed information that can aid in diagnosing the physical problems that are only partly reflected in the two health indicators. Similarly, diagnosing and treating mental health issues would require other diagnostics and treatments. Nevertheless, Specific Core Indices (partial performance indicators) are widely used, and can serve as a starting point for evaluating performance. Examples of specific core indicators for water utilities are listed in Appendix 1.

To illustrate the number and breadth of the performance indicators used within benchmarking initiatives, Figure 1 shows a matrix of PI categories developed by Canadian regulators as part of their Utility Management Tool. There are sixty-one Performance Measures for water utilities in the Canadian system—arguably too many for evaluating utilities (although the PIs include resource sustainability and environmental protection, two elements that traditional production and cost function studies omit). Even the twenty-three indicators for characterizing cost efficiency could be difficult to aggregate and/or evaluate. A production or cost model would be a useful complement to the specific indicators developed in the Canadian benchmarking initiatives.

Advantages of specific core indices: These partial measures provide the simplest way to perform comparisons; performance indicators are easy to calculate, they seem easy to interpret, and the data are often readily available from annual reports of companies. Focus can be given to trends over time to create targets for utilities.

Disadvantages of specific core indices: Such indicators fail to account for the relationships among the different factors. A firm that performs well on one measure may do poorly on another, while one company doing reasonably well on all measures may not be viewed as the "most efficient" company. Furthermore, an excessive number of specific indicators make it difficult to develop performance comparisons. Some sort of prioritization must be developed as well.

Application: Uganda's National Water and Sewerage Corporation has used a few Specific Core Indicators (unaccounted for water, collections, and network expansion) to create strong incentives for managers of local water utilities. Performance contracts based on achieving these targets involve managerial bonuses up to 50% of salary. Using simple trend information and comparisons across a set of utilities has had very positive impacts on NWSC performance.[1]

4.3 Aggregating Partial Indices into an Overall Performance Indicator (OPI)

The specific core indices are often combined in some way to create an overall performance indicator. The way this combination is usually performed is

[1] Silver Mugisha, Sanford V. Berg, and William T. Muhairwe (2007). "Using internal incentive contracts to improve water utility performance: the case of Uganda's NWSC," *Water Policy*, **9**(3), 271–284.

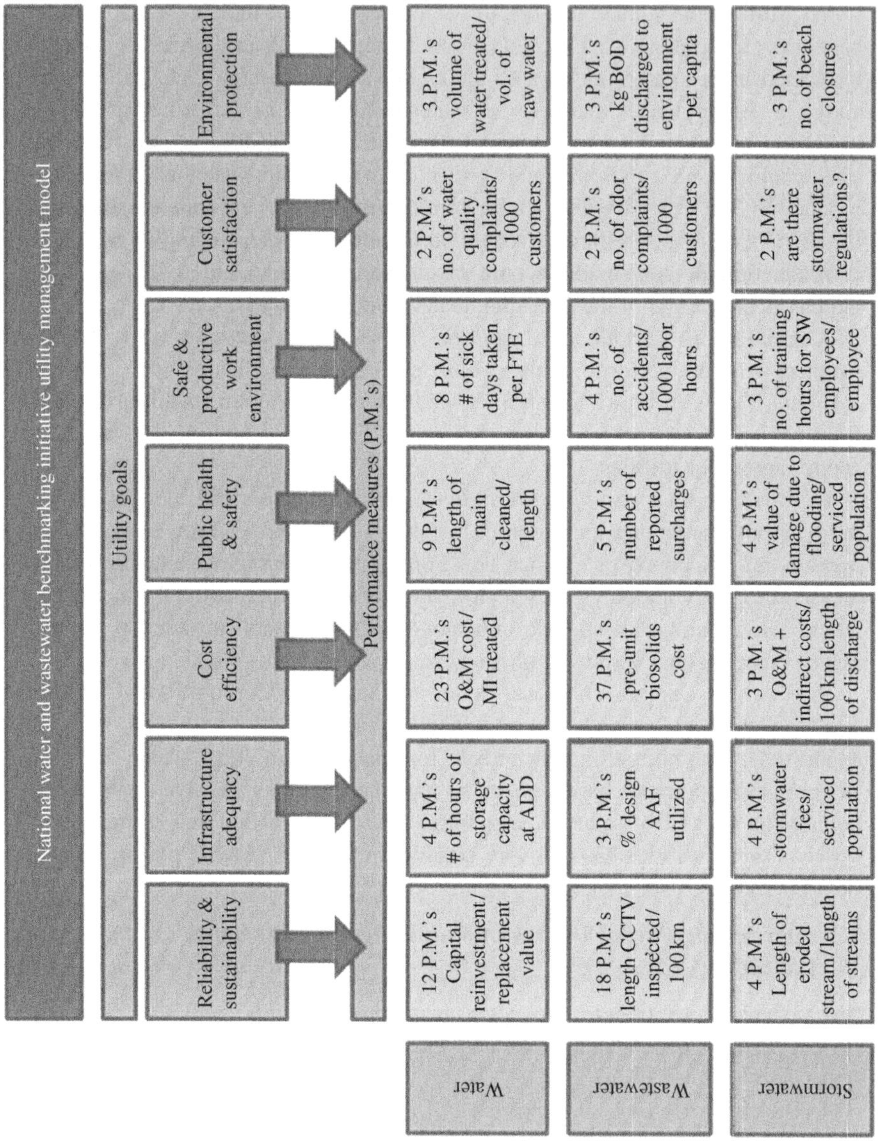

Figure 1. Goals and Performance Indicators

through a weighted average of core indices, where the weights reflect the importance the regulator assigns to each aspect of the firm performance.

Advantages of aggregating partial indices: The use of an OPI provides a summary index that can be used to communicate relative performance to a wide audience. Its components are easily understandable, providing some legitimacy to the resulting rankings.

Disadvantages of aggregating partial indices: The weights used to compute the OPI are generally not determined through a process that prioritizes different outcomes. Thus, when WSS are rated on the basis of a set of indicators, equal weight tends to be given to each component, even when the incremental costs of improving a particular ratio might be very high and the incremental benefits from the improved ratio minimal.

One problematic approach to aggregating Specific Core Indicators involves translating rankings into performance scores. In such a situation, a firm ranking 1, 3, and 5 in three categories would be given a score of 9, and a firm ranking 2, 2, and 8 would be given a score of 11. The former would "rank" higher than the latter. However, there is a problem when translating ordinal rankings into numbers, and summing those ordinal rankings as though they were scalars. Even assuming that equal weights are appropriate in the above case, the approach involves a scalar/ordinal problem. For example for the first two performance categories, the difference between the top four firms might be very small. Adding the ordinal rankings does not capture *relative performance* as reflected in the scalar scores. The second firm might have been bumped down because a few other firms happened to have higher scores on the third category; they were not "contenders", but pushed the second firm down in this particular ranking. So care must be taken when totaling the scores based on "ordinal rankings". Calculating the percentage of the best performer's score for each element (or performance relative to the "average") would yield an appropriate scalar; these could be totaled, though selection of indicators and determining weights remains an issue.

Application: The regulator for the water sector in Peru, SUNASS, utilizes a set of specific core indices to assess the relative efficiency of the service companies. SUNASS has selected nine indicators, grouped in the following four areas of efficiency:

(1) *Quality of service* groups three variables: compliance with the residual chlorine rule, continuity of service, and percentage of water receiving chemical treatment.
(2) *Coverage of service* groups two variables: water and sewerage coverage attained by each utility.

(3) *Management Efficiency* groups three variables: operating efficiency (a combination of continuity of service and the volume of water produced to serve each person in a connection), percentage of connections with meter installed, and the ratio of bills not yet collected to the total number of bills.
(4) *Managerial Finance Efficiency* is defined by the ratio of costs and expenses to revenues.

The first two areas of efficiency are intended to represent the interests of society. The third reflects the companies' performance, and the fourth represents the stockholders' perspective. Each indicator was assigned a weight of 1 as a first step in the benchmarking process. The emphasis on social fairness is evident in the greater number of indicators related to efficiency affecting society. Each indicator expressed as a percentage is multiplied by its weight and added to the percentage total per company. This total per company is divided by nine, the number of indicators. Finally, each company providing water and sanitation services is ranked on the basis of points attained within a group, with groupings determined by number of service connections, as follows:

- 17 are *small* with fewer than 10,000 connections.
- 20 are *medium*-sized, with 10,000–40,000 connections.
- 7 are *big*, with 40,000–160,000 connections.

SEDAPAL, the water and sanitation utility of Lima, has about one million connections. A comprehensive performance comparison would treat this large utility separately, since its inclusion in the sample would likely distort the results.

In Peru, each company sets its efficiency indicators within a master plan according to its financial and operational resources. This document is submitted annually to SUNASS for review and approval. Once the master plan is approved, it is viewed as involving mandatory targets. At the end of the year, the ranking based on actual data is calculated by SUNASS and published.[2]

Similarly, the SEAWUN Benchmarking program in Asia develops an "Overall Performance Indicator" (OPI) based on the ten categories in the IBNET system: coverage, water consumption and production, unaccounted for water, metering practices, pipe network performance, costs and staffing,

[2] For more on Peru, see Maria Luisa Corton (2003), Benchmarking in the Latin American water sector: The case of Peru. *Utilities Policy*, **11**, 133–142.

quality of service, billings and collections, financial performance, and capital investment. The OPI reflects relative performance on fourteen partial indictors.

4.4 Performance Scores Based on Production and Cost Estimates ("Total" Methods)

The main argument of this volume is that production and cost metrics should be included in the toolkits of decision-makers who evaluate water utility performance. These methodologies have been discussed in some detail. But it is useful to introduce additional observations about these techniques. Total methods capture the interrelationship analysis of *several inputs and outputs* of a single firm, allowing for the identification of distinct types of efficiency/inefficiency such as technical or cost efficiency. Total methods may be classified in terms of the primary intent of the benchmarking exercise, the type and quantity of data available, and the resources at the disposal for the needed calculations. As with overall performance indicators, cross sectional comparisons within an industry or with other countries may complement or enrich the analysis. In particular, international benchmarking is used when there are not enough data, as in the case of a newly reformed sector or in the presence of a national monopoly. However, when making international comparisons, caution is necessary; analysts must account for those factors that differentiate the countries being analyzed. Differences include economic, political, topological, meteorological and technological factors.

In a survey of energy regulators from forty nations, Haney and Pollitt find that benchmarking techniques are widely used in the regulation of gas and electricity networks. However, few apply best practice techniques "respondents pointed to a number of limitations of advanced benchmarking techniques, particularly in terms of data requirements, comparability and interpretation difficulties."[3] These same issues face water regulators—as they attempt to develop valid comparisons of performance across firms.

4.4.1 Index methods (Total Factor Productivity)

Concept: How much output is achieved with each unit of input? Total Factor Productivity is a measure of the physical output produced from the use of a given quantity of inputs by the firm. When having multiple outputs and multiple inputs, the ratio of the weighted sum of outputs with respect to the weighted sum

[3] Aoife Brophy Haney and Michael G. Pollitt (forthcoming), *Energy Policy*, "Efficiency analysis of energy networks: an international survey of regulators."

of inputs is used to calculate the Total Factor Productivity Index. In general, the weights are the cost share for inputs and the revenue shares for the outputs.

Prices or cost shares and revenue shares may change between two periods. There are two alternatives in dealing with this problem which implies different calculations: same weights may be used in both periods, or each period may use a different weight. The focus of the analysis should be the factor by which the input in one period could be decreased such that the firm could still produce the same output level of a comparing period. This approach yields the Malmquist index. In this case, the choices for calculation are the use of the final period, initial period, or geometric mean of both periods as the reference point. The Malmquist index may be used either with parametric or nonparametric specifications, but is more generally associated with DEA.

When data on input quantities are not available, an indirect TFP index may be used. We could deflate total revenue and total costs by a suitable price index to obtain a quantity index. Then, we could directly use deflated revenues over deflated costs. If the price index does not have the same scope (e.g., the input price index is related to the economy in general and the output price index is related to the particular industry), the TFP index will be biased. Furthermore, when measuring the TFP index change for a group of firms, pair wise comparisons will not necessarily give the same rankings of firms. These indexes need to be adjusted using a more complex calculation technique.

Advantages of index methods: The approach only requires two observations, such as data on two firms or for one firm from two time periods.

Disadvantages of index methods: TFP cannot be decomposed into the different types of efficiencies (i.e. technical, allocative, and economic, as mentioned earlier). These refinements may not be important if the evaluator is only trying to gauge broad trends.

Application of index methods: This method has been used in the USA telecommunications sector, where the FCC has compared industry TFP with that for the economy as a whole.

4.4.2 Estimation using mean and average methods

Section III provided graphical examples of "fitting" observations to a linear model. The basic statistical skills required for performing such tests are not substantial, but care must be taken about model specification and data availability. Extreme values observations ("outliers") can unduly affect the estimated coefficients, and thus performance evaluations of specific WSS. There are basically two methods within this category: Ordinary Least Squares (OLS) and Corrected Ordinary Least Squares (COLS):

Ordinary least squares (OLS)

Concept: OLS techniques can be used to perform benchmarking that relates individual firm performance relative to what would be expected: an estimate of an average production or cost function of a sample of firms. Average benchmarking methods may be used to compare firms that face similar conditions, and therefore have relatively similar costs or when there is a lack of sufficient data of comparable firms for the application of frontier methods. Basically, the method refers to the estimation of a regression functional form for costs or production models using the OLS approach. Regression analysis seeks to derive a relationship between firm performance (in terms of output or total cost), and market conditions, including characteristics of the production processes. Statistical analysis can isolate the impacts of specific conditions or levels of output—so the roles of multiple independent variables can be determined. Data from the firms being compared can then be used to arrive at expected dimensions of firm performance, given the variables characterizing each firm.

The technique of regression analysis is defined by the following steps: 1) selecting both the cost (or output) measure and exogenous variables, 2) estimating a cost (or production) function for the industry, and 3) calculating the efficiency coefficient for each firm within the industry. Predicted versus actual output provides a measure of relative performance. The quality of these results can then be statistically evaluated to provide the policy-maker with a framework for evaluating firms.

Corrected ordinary least squares (COLS)

Concept: A slightly different approach than OLS involves shifting the regression line (or hyper-plane when there are many independent variables) towards the best performing company, which is called Corrected Least Squares methodology (COLS). In a general sense, COLS is merely a shifted average function. Two steps are needed, one to get the expected value of the error term and another to shift or to "center" the equation.

When using OLS or COLS it is good practice to perform Quantile analysis, which considers points at regular intervals within a cumulative distribution function. Quantile analysis helps to overcome the possible effect of outliers on the estimated mean allowing the analyst to detect the presence of performers on specific or extreme quartiles such as the lower (25%) or the upper (75%) quartiles—the term used when the distribution is divided into four parts.

Advantages of mean and average methods: The statistical method reveals information about production functions or cost structures and identifies the differential impacts of key variables on output. Specific coefficients can be estimating, including those representing the degree of scale economics.

Disadvantages of mean and average methods: A large data set is necessary in order to obtain reliable results. The regression results are sensitive to functional form if the error term is not adequately specified, which can lead to varying interpretations depending on model specification.

Application: The UK water regulator OFWAT applies mean and average methods to the operating costs (OPEX) and capital expenditures (CAPEX) of water utilities when determining the price caps every five years. OFWAT has developed an efficiency analysis relying on mean and average methods that is a key part of its price determination process. OFWAT argues that:[4]

> "The price limits we set companies are based on demanding efficiency assumptions. This encourages companies which perform poorly to catch up with the more efficient ones. We make comparisons between companies, against each individual company's targets, and with other sectors. We ask them to improve if this is necessary. In a monopoly of this sort we rely on comparative competition. We give companies incentives to keep their running costs and the costs of maintaining their assets to the minimum necessary and to use cost-efficient solutions to achieve their outputs. Where companies are successful, they retain the savings for five years and customers benefit at the next price review through price limits which take account of the efficiencies achieved."

The application of price caps (along with quality standards) for privately operated water utilities certainly places pressure on managers for cost containment. Weak productivity performance relative to others can be used by the regulator to set lower prices, arguing that the firm has significant opportunities to reduce costs (since comparable firms have demonstrated greater efficiencies).

4.4.3 Frontier methods

Concept: Frontier Benchmarking places a greater emphasis on performance variations relative to the top performing firms than does OLS. Frontier-based benchmarking methods estimate (using regression techniques) or identify (using DEA techniques) the efficient performance frontier for a sample of firms. The efficient frontier is the benchmark against which the relative performance of firms is measured. Given a particular sample, all companies are assumed to be able to operate at an optimal efficiency level which is determined by the most efficient companies in the sample. These efficient companies are sometimes

[4] See http://www.ofwat.gov.uk/aptrix/ofwat/publish.nsf/Content/protecting_interests 280905

referred to as the "peer firms" and determine the efficiency frontier. The companies that form the efficient frontier use minimum quantity of inputs to produce the same quantity of outputs. The distance to the efficiency frontier provides a measure of inefficiency. If a large data set is available, this approach may be suitable at the initial stages of regulatory reform when the most pressing objective is to reduce the performance gap within the industry. Both approaches are now explained in a little more detail:

Stochastic frontier analysis: Stochastic Frontier Analysis attempts to empirically estimate an efficient frontier, incorporating the possibility of measurement error (or random factors) in its estimation. To separate inefficiency and noise, the statistical analyst needs to make strong assumptions regarding the distribution of error terms within the sample. Stochastic frontiers may be further classified into *Production*, *Cost*, and *Input Distance* frontiers.

A ***production frontier*** reveals the relationship between inputs and outputs of firms and represents an alternative when cost frontiers cannot be calculated due to lack of data. The estimated output is the maximum possible output for given inputs of an individual firm. The output difference obtained in the estimation is interpreted as technical inefficiency of each individual firm. A production frontier allows explicit consideration of variable returns to scale: appropriate scale efficiency changes need to be included when calculating total factor productivity.

A ***cost frontier*** shows costs as a function of the level of output and input prices. It is useful when trying to determine output prices that would allow utilities to recover "efficient" costs (not just realized costs). Conceptually, the minimum cost function defines a frontier showing costs associated with various levels of inputs, the input prices, and other control variables. A cost function in two dimensions would show how Cost (on the vertical axis) depended on the output (shown on the horizontal axis). Ideally, analysts should attempt to identify the Total Cost frontier rather than variable cost frontier to recognize the substitutability among factor inputs. However, determining an annualized capital cost (separate from capital expenditures in a particular year—CAPEX) is not an easy task, since asset valuation and determining the cost of capital are both difficult tasks, especially when evaluating state-owned enterprises. A number of studies just attempt to estimate short run cost frontiers using OPEX. Cost efficiency reflects the effects of both technical and allocative efficiency.

The production and cost approaches can yield different efficiency scores and efficiency rankings. The difference will be larger if large allocative distortions are present, since inappropriate input mixes are utilized. In this case, the parameters of the cost frontier will be biased. An important factor to consider when choosing between a cost frontier and a production frontier is that usually regulated firms are required to provide the service at a preset price and the utility must meet demand. In this sense, firms are not allowed to choose their own level of output, which makes output an exogenous variable. The regulated firm maximizes benefits by minimizing its costs of producing a given level of output, so a cost frontier approach is appropriate for making performance comparisons.

Finally, an ***input distance frontier*** is another option for analyzing regulated industries: output quantity is exogenous (determined external to the utility) and input quantities are endogenous (determined by the utility), and when the technology involves multiple outputs or there is insufficient data on input prices. This situation characterizes a water and sewerage utility which produces several outputs and where some inputs may be shared.

A distance function may have either an input or an output orientation. An input orientation looks at how much the input vector may be proportionally contracted with the output vector held fixed. An output orientation looks at how much the output vector may be proportionally expanded with the input vector held fixed. Input distance functions can be estimated by either stochastic or DEA methods. The advantage of a distance frontier with regard to a cost frontier is that firm is not assumed to be minimizing costs. An advantage of the production frontier is the avoidance of endogeneity.

Advantages of stochastic frontiers: These models account for data noise such as data errors and omitted variables. Standard statistical tests can be used to test hypotheses on model specification and significance of the variables included on the model. It is also more amenable to modeling effects of other variables (e.g., environment, quality).

Disadvantages of stochastic frontiers: Analysts need to specify functional form and production technology. Also, the separation of noise and inefficiency relies on strong assumptions regarding the distribution of the error term. The statistical skills are more demanding for advanced econometric techniques.

Application: A number of studies utilize SFA, as evidenced in Appendix 2. The selection of a Cost, a Production or a Distance function depends on particular circumstances regarding five areas:

(a) *Operating environment*: Assuming that the firm is minimizing costs (or, maximizing profits), a production function or a distance function may be a natural option. If instead the firm has an "obligation" to service, as when it is required to supply full demand, a cost function may be more suitable for making comparisons.
(b) *Possible endogeneity*: It is been argued that inputs in a production function are jointly determined with the output produced. This leads to an econometric issue: endogeneity which may produce inconsistent and biased parameters estimates. Obviously, this issue may be addressed

using certain econometric tools; however, the estimation procedure becomes more complicated when not enough variables are available. Cost or distance function specifications do not present this problem.

(c) *Data limitations*: The main issue when using a cost function is the availability of prices of inputs, in particular the price (cost) of capital. Variables such as cost of energy, and labor are seldom reported by firms. This limits the use of a cost function approach favoring either a production or a distance function.

(d) *Output definition*: From a production process point of view, a firm may produce more than one marketable product as with many utilities in the water sector. Econometric analysis using a production function will entail the specification of several equations, which makes the estimation process more complex. When having multiple outputs, the analyst will use either cost or distance functions. In addition, DEA can also be utilized.

(e) *Technology specification*: It is well known that factors other than input quantities may have an effect on the production process: such factors include the economy, topology of the location, age of the network, or even company ownership. All these characteristics have implications for the production process. There are two ways to introduce these factors into a production function model: by assuming that they directly affect the production process (technology) or by considering them as elements affecting the firm's efficiency. Introducing these variables into a cost function specification requires more care since it may not be clear how they affect the cost structure directly. Thus, it is necessary to carefully address these issues before model specification.

Non-stochastic frontiers: Data envelopment analysis (DEA): As has been noted, DEA is a method in which linear programming techniques are applied to a framework of a multi-input, multi-output production function. DEA benchmarks firms only against the best producers. The approach can be characterized as an extreme point method. The assumption is that if any one firm can produce a certain level of output utilizing a specific number of inputs, then any comparable firm of equal scale should be capable of doing the same. The most efficient producers for any given level are used to form composite (or hypothetical) producers, allowing the analyst to identify an efficient solution for every output level.[5]

[5] A useful website at Warwick University contains tutorials and examples on DEA: http://www.deazone.com/.

If the analysis uses cost, DEA searches for the points with the lowest unit cost for any given output, connecting those points to form the efficiency frontier. Any company above this minimum total cost frontier is identified as inefficient. A numerical score is given to each firm, such as the efficient cost divided by the actual cost at that output level.

Advantages of DEA: The method can easily accommodate a multiplicity of inputs and outputs, if the number of observations is adequate.

Disadvantages of DEA: The results are sensitive to the selection of inputs and outputs, so their relative importance needs to be determined prior to the calculating an efficiency score. However, there is no way to test for the best specification. The number of efficient firms on the frontier tends to increase with the number of inputs and output variables.

Application: For evaluating firms in the electricity sector, the Norwegian authorities chose Data Envelope Analysis as their measuring tool to design an individualized regulatory rate for each firm based on their comparative efficiency. The input components included man-hours, losses, capital (lines and transformers only), and goods and services. The output variables were number of customers, energy delivered, length of lines, and degree of coastal exposure.

4.5 Examples of Empirical Studies

Analysts have applied benchmarking techniques to infrastructure firms. Some of these studies go beyond technical discussions of particular parameters and proceed to make performance comparisons among firms in the sample. There are far more quantitative analyses of electricity systems than water systems. For example, less than a decade ago, Jamasb and Pollitt (2001) surveyed a large number of studies examining scale economies and economies of scope for electricity suppliers.[6] Recent surveys of water studies have identified the increase in cost and production function studies for the water sector. Earlier, this volume noted that Graham Shuttleworth is very critical of studies that try to identify the relative performance of firms. Among other concerns, he argues that (for electric utilities):

> "Our experience in Europe shows, however, that different versions of a DEA model will give quite different results and there is no way to tell which set of results is most reliable. Picking the "right" set of DEA scores then becomes just a matter of opinion, based on how inefficient one thinks (or would like to think) a company is."

[6] T. Jamasb and M. Pollitt (2001). "Benchmarking and regulation: International electricity experience, *Utilities Policy*, **9**(3), 107–130.

Such skepticism is not shared by all analysts, but Shuttleworth's comments serve as a useful antidote to anyone who believes that running statistical regressions or DEA models will lead to easy answers regarding relative performance. Most of his concerns have been mentioned earlier when explaining the reasons for not engaging in benchmarking.

Clearly, robust rankings will require that several methodologies be utilized, and differences explained. If the same set of firms has relatively high scores (or low scores) regardless of methodology and over a range of reasonable model specifications, we can have some confidence that the procedures have identified strong (and weak) performers. Turning to water systems research, Table 1 presents a listing of eight empirical studies of water systems. The table illustrates the wide range of studies that have been conducted over the past two decades.[7] These studies are not meant to be the *Top Eight* in number of citations, technical virtuosity, or policy impact. Rather, they represent different types of analyses covering a wide range of nations. The advantage of academic studies is that the authors generally do not ask those paying for the study: "What conclusions do you want us to support?" Rather, the studies are likely to be objective; the authors' reputations are on the line and the empirical work has to pass external reviews. At the same time, they are less likely to be easily utilized by managers or regulators. Of course, consultants also have reputations to maintain, so the best consultant studies are carefully done, use appropriate quantitative techniques, and must be consistent with their past work.

The listing illustrates the scope of studies on water utility costs and production patterns. Academic journals tend not accept articles on the basis of their decision-relevance (such as providing specific efficiency rankings of utilities in a nation and the associated confidence intervals around each "score"), but on how innovative the study is and the insights emerging from the analysis. Within the university community, technical virtuosity tends to trump usefulness, except in the most applied journals. To some extent, this is an appropriate division of labor: consultants and organizations can perform studies that help regulators evaluate relative performance.

So academics and consultants play different roles in the regulatory process. Nevertheless, one purpose of specialized research is to provide new tools ("fire-fighting equipment") for decision-makers (those fighting fires). This means that academics have some responsibility for making new techniques accessible to those who are attempting to quantify utility performance. Thus, a number of

[7] The author (with with Rui Marques) is preparing a meta-analysis of over one-hundred articles published in scholarly journals. Thus, a comprehensive literature survey is beyond the scope of the present volume.

Table 1. Illustrative Empirical Studies of Water Systems

Peru: Corton, Maria Luisa (2003). "Benchmarking in the Latin American water sector: The case of Peru," *Utilities Policy*, **11**, 133–142.
Data: 44 water utilities in 1998.
Methodology: Ordinary Least Squares OPEX = f (volume of water produced, network size, and other variables).
Question: What are the operating cost drivers in Peru? What is the relative efficiency of the different water utilities in Peru, controlling for factors beyond their control?

Brazil: H. Tupper and M. Resende (2004). "Efficiency and regulatory issues in the Brazilian water and sewage sector: an empirical study," *Utilities Policy*, **12**, 29–40.
Data: 25 state utilities during 1996–2000.
Methodology: Data Envelopment Analysis (controlling for network density and water losses).
Question: What is the relative efficiency of water and sewerage utilities in this sample of utilities? What kinds of incentives can be based on benchmarking studies and how can incentives be implemented so as to improve sector performance?

England and Wales: D. Saal and D. Parker (2006). "Assessing the performance of water operations in the English and Welsh water industry: a lesson in the implications of inappropriately assuming a common frontier," in T.J. Coelli and D. Lawrence (eds.), *Performance Measurement and Regulation of Network Utilities*, Edward Elgar, Publisher, pp. 297–327.
Data: 180 observations for water-only utilities and 110 observations for water and sewerage utilities during 1993–2003.
Methodology: Quality-adjusted input distance function and Stochastic Frontier Analysis, controlling for density (connections per kilometer of water mains), average pumping head (lift times the volume of water), and water quality (percentage of water in compliance with nine key water parameters).
Question: Can a common frontier be utilized for both water-only and water and sewerage companies? What is the rate of technological change for water-only utilities?

Australia: Coelli, Tim and Shannon Walding (2006). "Performance measurement in the Australian water supply industry: a preliminary analysis," in T.J. Coelli and D. Lawrence (eds.), *Performance Measurement and Regulation of Network Utilities*, Edward Elgar, Publisher, pp. 29–55.
Data: 18 water supply utilities for 1995/96–2002/03.
Methodology: Data Envelopment Analysis with two outputs (connections and volume of water delivered) andtwo inputs (OPEX and capital).
Question: What are the efficiency scores of utilities? What is the rate of total factor productivity change for water supply utilities? How could the results be utilized to select X-factors in a price review?

(*continued*)

Table 1. Continued

Italy: B. Antonioli and M. Filippini (2001). The use of a variable cost function in the regulation of the Italian water industry, *Utilities Policy*, **10**, 181–187.
Data: 31 water distribution utilities from 1991–1995.
Methodology: Variable Cost as a function of output, price of labor, size of the water system (length of network), number of customers, a proxy for capital stock (number of water wells), water losses and a zero-one dummy variable capturing the need to treat source water to meet health standards—using a Cobb Douglas cost function.
Question: Are there economies of density and of scale in Italian water utilities? How does the incorporation of service area characteristics for water utilities affect the estimates of relative efficiency? How can robust benchmarking methods be utilized by regulators?

Uganda: Mugisha, Silver (2007). "Performance assessment and monitoring of water infrastructure: an empirical case study of benchmarking in Uganda," *Water Policy*, **9**(5), 475–491.
Data: 1996–2004, from 8 to 14 utilities.
Methodology: Stochastic Frontier Analysis using a Translog production function.
Question: Has managerial orientation (from engineering vs. customer-driven) affected the efficiency of local water operations?

Cross Country: C. Nauges and C. van den Berg (2008). "Economies of density, scale and scope in the water supply and sewerage sector: a study of four developing and transition economies," *Journal of Regulatory Economics*, **34**, 144–163.
Data: 154 water utilities in Brazil, Moldova, Romania and Vietnam, where the years range from 1996–2004.
Methodology: Translog Variable Cost Function that includes time and utility-specific effects, using a system of seemingly unrelated equations.
Question: Are there economies of density, scale and scope for water supply and sewerage? Do these differ across countries which are at different stages of development?

Cross-Country in a Region (Asia): A. Estache and M. Rossi (2002). How different is the efficiency of public and private water companies in Asia? *The World Bank Economic Review*, **16**(1), 139–148.
Data: 50 water companies in 29 Asian and Pacific region countries for 1995.
Methodology: stochastic cost frontier Cobb-Douglas
Question: Are there differences in efficiency between public and private utilities?

journals attempt to bridge principles and practice—but technical elements of the research still tend to be given high priority in the review process.

These empirical studies are briefly summarized in Appendix 2: Annotated Bibliography of Benchmarking Studies. As one can see from the publication dates, there has been significant activity in benchmarking in recent years. Consistent data covering a number of years and firms are becoming available for a number of nations. However, most of the research has been academic in nature, or has involved external consulting contracts. Transferring requisite technical skills to regulatory commissions and companies will be essential if such comparisons are going to influence incentives in the future.

Data availability: Although a cost function model is the preferred choice when analyzing cost efficiency, the use of total cost or variable cost functions is also framed by availability of data. When assuming the stock of capital as fixed, meaning that elements of the firm's machinery or equipment are not easily altered in a certain time period, the use of a Variable Cost specification is indicated and stock of capital enters as explanatory variable in the a variable cost function. This is a sensible assumption for the water sector where assets have very long service lives. Implicitly, it underscores the fact that current management does not have influence over fixed factors such as capital.

However, the possibility of substantial input factor substitution among capital and other input factors, in particular labor, may suggest that a Total Cost function model is more appropriate. Yet, the limitation here is, as it has already pointed out, data availability.

Functional form: Either for a Cost, Distance or Production approach, the functional form requires careful attention. We have seen how linear and non-linear relationships might be captured in an empirical analysis. Appendix 1 provides definitions of variables that have been used in benchmarking studies:

- Output Variables
- Quality Variables
- Input Variables: Quantities and Prices
- Accounting/Financial Variables
- Conditioning/Environmental Variables
- Macroeconomic Variables
- Governance Structure Variables

Appendix 2 contains an Annotated Bibliography of Water Benchmarking Studies. The reader can see the wide range of questions being asked: extent of scale economies, degree of economies of scope between water and sewerage production, and the relative efficiency of different water utilities, to note just a few. As noted earlier, the analyst should utilize several methodologies and specifications to test for the sensitivity of results to the number of years, the cross-section sample, variables, and model specification. Data availability and the issue under investigation will affect model specification. Statistical tests described in Appendix 3 can be used to determine which functional form is most appropriate.

Other methodologies: Several other methodologies were mentioned earlier. The *Engineering/Model Company* approach has been used to establish baseline performance expectations in a number of countries, including Chile and Brazil. The use of an "artificial" firm that has optimized its network design and minimizes its operating costs can provide insight into what is possible if a firm is starting as a Greenfield Project. As with the metric methodologies described in this volume, the approach also has its limitations. A second (and relatively popular) methodology focuses on individual production processes: *Process Benchmarking*. One advantage of this approach is the ability to identify specific stages of the production process that warrant attention if WSS performance is to be improved. A third methodology, *Customer Survey Benchmarking*, focuses on the perceptions of customers as a key element for performance evaluation. Each approach has strengths and weaknesses; this technique can shed light on consumer concerns, reflected in complaints or in customer surveys. *Financial sustainability Benchmarking* and *Water Resource Sustainability Benchmarking* are also relevant for evaluating the long term effectiveness of current cost and revenue structures. As in medical evaluations of people, no single index captures the complexity of an organization's health.

Chapter 5 Summary and Conclusions

This volume provides material that complements other IWA books, IBNET, World Bank Toolkits and the growing set of resources becoming available to those developing and implementing public policy in the water sector.[1] The purpose is to help decision-makers become more effective producers *and* consumers of benchmarking studies. Such studies improve prospects for high performance and increase the likelihood that resources will be used effectively in meeting citizen needs. We have seen how benchmarking specialists using cost and production functions prepare and critique studies that utilize various quantitative methodologies. We know that rankings can be manipulated by choice of variables, model specification, sample size, time frame, and treatment of outliers. Because the stakes are high, studies can be controversial: affected parties have an interest in the relative and absolute performance evaluations prepared by analysts.

Economist Joan Robinson offered a view on the benefits of applying economic concepts to policy issues:

> *The purpose of studying economics is not to acquire a set of ready-made answers to economic questions, but to learn how to avoid being deceived by economists.*

A counter-point to Robinson's somewhat cynical view (or realistic view, depending on your attitude towards economists) is provided by John Maynard Keynes:

> *The body of economics does not form a body of settled conclusions immediately applicable to policy. It is a method rather than a doctrine, an apparatus of the mind, a technique of thinking, which helps its possessor to draw correct conclusions.*

Both observations are correct: Figures lie and liars figure, but there are techniques that allow us to move closer to the truth. There is strong evidence that

---

[1] Benchmarking Water and Sanitation Utilities: A Start-Up Kit. May 1999. World Bank. Washington, DC.

cost and production studies can lead to rankings that approximate reality. Clearly, professionals in organizations funding and conducting studies need to understand the strengths and limitations of metric benchmarking.

5.1 Potential Impacts of Benchmarking Studies

While benchmarking studies are no panacea for improving utility operations or overcoming impediments to public and private investment, they do provide key inputs for public policy discussions and managerial evaluations. Citizens will remain unaware of what is possible if they have no basis for making comparisons. Managers cannot manage if they are not collecting and evaluating operating and financial data. Metric benchmarking can identify outliers—providing policy-makers and managers guidance on where to improve. The benchmarking of activities and services allows decision-makers to focus on what to improve. Finally, process benchmarking facilitates the sharing of best practice techniques—helping managers determine how to improve. Thus, benchmarking studies have implications for network expansion, poverty reduction, organizational incentives, and international and national support for new initiatives.[2]

5.1.1 Network expansion

Network expansion requires capital which can only come from five sources: operating cost reductions, customer price increases, government budgets (taxpayers), private investors (via equity ownership or bonds), or international organizations.[3] Financial and operating data are essential for those evaluating whether investment funds are likely to be used effectively. When a utility has no track record, national and international investors will view associated projects as relatively risky, signaling a reduced likelihood of private participation. National governments also have alternative uses of funds: education and health compete with other infrastructure sectors during the budget process. If managers cannot demonstrate good stewardship of public funds in the past, they are unlikely to be perceived as strong candidates for further funding

---

[2] The political economy of the water sector underscores the importance of greater transparency and benchmarking. See Matthias Krause (2009). *The Political Economy of Water and Sanitation*, New York: Routledge, xix–252.

[3] Typically, investors seek government guarantees, but guarantees can blunt incentives to select efficient projects. In addition, guarantees can become liabilities affecting government budgets. For an overview of public and private initiatives, see William Easterly and Luis Serven (2003), *The Limits of Stabilization: Infrastructure, Public Deficits and Growth in Latin America*, World Bank, xv–208.

(whether via budget appropriations or investments by national development banks or pension funds). Those with fiduciary responsibility will require that utilities demonstrate that performance trends are in the right direction. Furthermore, political entrepreneurs who might support greater investment in water and wastewater infrastructure will want clear justification for the use of public funds in this way. In addition, multinational organizations like the World Bank will only be willing to support water system investments if studies document the effectiveness of past initiatives and the likely success of new programs.

5.1.2 Poverty reduction

Lack of citizen access to clean water and public perceptions regarding infrastructure reforms affect voter confidence in the political process.[4] Justice might be characterized in two ways: fair processes and fair outcomes. Fair procedures yield reliable information, promote consistency in decisions, and take into account the viewpoints of affected parties. Decisions reflecting transparent processes (including the publication of benchmarking comparisons) will promote public acceptance of outcomes and contribute to the legitimacy of social arrangements. When water is viewed as a commodity whose provision requires investments and operating outlays, procedural fairness will emphasize the financial sustainability of arrangements. When water is viewed as a human right, outcomes related to universal access are emphasized; principles of distributive justice tend to take precedence under when water is viewed as a human right. Both notions of fairness are relevant when evaluating new infrastructure initiatives. Benchmarking contributes to increased efficiency, which in turn promotes distributive justice to the extent that resource savings are applied to projects benefiting the poor and disadvantaged. Note that political rhetoric in support of universal service is not always accompanied by significant increases in public investments that expand networks and protect water sources. From this perspective, some calls for distributive justice do not even meet the standard of compassion, as when resources are not allocated by the state to meet basic human needs or are wasted through political patronage and corruption.[5]

---

[4] *Water Governance for Poverty Reduction: Key Issues and the UNDP Response to Millennium Development Goals* (2004), United Nations Development Programme, 1–93.

[5] These points are extracted from S. Berg and S. Mugisha (forthcoming) "Pro-Poor Water Service Strategies in Developing Countries: Promoting Justice in Uganda's Urban Project" *Water Policy.*

5.1.3 Organizational incentives

Benchmarking is central to the development and implementation of incentives for improving WSS service performance.[6] Here, we can consider two types of incentives, those that are externally established and those associated with internal management processes. A sector regulator often has authority to set prices. For a well-performing privately-owned utility, the potential for further cost reductions may be minimal so the regulator may allow prices to increase at the rate of inflation. A utility with a very low efficiency score has greater scope for cost containment, so its prices might be set so they decline over time—putting pressure on management to improve performance. Benchmarking comparisons make such regulatory policies feasible. An agency that monitors a management performance contract may not be in a position to establish incentives, but its role is to enforce the contract—ensuring that the terms and conditions are, indeed, met by all parties. Generally, such contracts include penalties if targets for quality of service or targets for network coverage are not met by specified dates. Similarly, if a municipality operates the water utility, there should be some mechanism for the City Council to review budgets and oversee performance. In all these cases, benchmarking reduces the information asymmetry between managers and those providing oversight. Well-governed utilities will have incentive programs for managers and staff to meet performance targets. Again, the process requires management to document strong and weak performers (whether by geographic district or different stages of production).

5.1.4 International support and national commitment

Improved performance within the water sector is one of the Millennium Development Goals. Yet we know that many nations will not achieve the targets for universal service expansion. Health will suffer and citizens will feel abandoned by the national governments and international agencies that signed on to these goals. Over the past three decades, many billions of dollars have been transferred from developed to developing countries, yet the results for water and wastewater systems (in terms of operations, service quality, and network expansion) are quite mixed. Future initiatives will come under greater scrutiny, since the donor funds are limited; multilateral guarantees will depend

[6] Mehta Meera (2003), *Meeting the Financing Challenge for Water Supply and Sanitation: Incentives to Promote Reforms, Leverage Resources, and Improve Targeting*, World Bank, May, 1–136.

on the credibility of business plans. However, there is renewed interest in supporting projects involving utilities with strong track records. Benchmarking is one tool that can demonstrate organizational capabilities to use additional funds wisely. Developed nations face the same resource constraints. Estimates of needed investments in system renewal, network security, and energy efficiency run into the hundreds of billions of dollars. Just subsidizing utilities from national tax revenue is not a good use of public funds. Prioritizing support funds requires benchmarking studies: decision-makers need evidence regarding who will use funds wisely. Furthermore, national support in OECD countries should be in the form of loans not subsidies, so customers in wealthy nations begin to pay the full cost of service. In the future, both non-OECD and OECD nations will be paying greater attention to the efficient use of funds in the water sector.

5.2 Concluding Observations

This examination of measurement issues and methodological alternatives has implications for establishing performance incentives. Managers of high-performing utilities merit rewards and those managing utilities with low scores (or poor efficiency trends) need to have additional training or be replaced. This volume attempts to make concepts used by economists accessible to those engaged in reforming, analyzing, regulating, and managing water and sewerage service utilities. To date, data specification, collection, and collation has been the focus of benchmarking programs.[7] With several years of consistent data available for many utilities, the issue becomes one of how to utilize the data (including core indicators) in evaluating and rewarding utility performance. Since the technical literature is vast (and often inaccessible to all but the most technically proficient individuals), this volume fills a gap that hopefully will help practitioners initiate more comprehensive analyses with these data.[8]

The author has tried to highlight the strengths and limitations of alternative methodologies without becoming excessively technical. Benchmarking studies

---

[7] "Review of the water supply and sanitation sector benchmarking practices and models in the EU and other regions", Report for 6th EU Framework Programme for Research and Technological Development. March 1, 2004.

[8] The World Bank Institute has published a technical volume that serves as a good starting point. See Coelli *et al.* (2003), *A Primer on Efficiency Measurement for Utilities and Transport Regulators* (WBI Development Studies), xii–134.

directed at practitioners have tended to focus on Core Indices and Summary Indicators rather than the models and methodologies associated with production and cost functions. This volume emphasizes three tasks for those who would improve performance in the water/wastewater sector: (1) making performance comparisons in a systematic way, (2) using the performance scores to create incentives, and (3) promoting acceptance of the process.

5.2.1 Comparing performance

A wide variety of methodologies are available for evaluating performance. Here the focus has been on tools used by economists, particularly techniques that determine the extent to which utilities are on the production and/or cost frontier. Benchmarking can help decision-makers identify strong and weak performance improve, including meeting service quality standards. Any benchmarking study will have limitations, but sound studies can be used to place the burden of proof on other parties who might argue that the analysis is incomplete or incorrect. Over time, as professionals gain experience with these quantitative techniques, data availability will improve and studies will be strengthened. In the process, governance procedures within companies can incorporate this information into managerial incentive packages. Thus, rankings can serve as catalysts for better stewardship of water and other resources. Still, care must be taken to use comprehensive indicators, lest those being evaluated "game" the system. If only a subset is used, performance may improve for some dimensions of a firm's operations but may become worse for others.

Benchmarking studies should target specific audiences with appropriate levels of detail. Each of the five benchmarking approaches described here can provide insights on aspects of water utility performance. For example, *Overall Performance Indicators* provide information time trends and patterns across firms. This volume focused on *Performance Scores based on Production or Cost Estimates* ("total" metric methods); these techniques utilize the data sets that are becoming more widely available. Regional and international initiatives have been consolidated under the auspices of IBNET—leading to comparable data over time and across a large number of water utilities. Even prior to IBNET, research on production and cost functions grew at a rapid pace: nearly two hundred empirical studies have been identified to date, and new studies appear each year. Techniques that are being applied require a variety of technical skills. Care must be taken when interpreting the results of quantitative studies. However, these methodologies can provide important information for analysts who assess industry and firm performance.

This volume has identified the steps necessary to create a successful benchmarking program:[9]

- *Create a benchmarking team*: professionals need a variety of technical and engineering skills so the Team has a comprehensive understanding of the water utility's processes and operations;
- *Establish clear study objectives*: the key questions frame the issues at hand and identify data needs;
- *Select an appropriate benchmarking methodology*: analysts must be aware of the strengths and limitations of techniques for analyzing cost and production functions;
- *Gather raw data for yardstick comparisons*: this stage involves the selection of the appropriate comparison group of companies or standard (depending on the benchmarking method selected);
- *Apply data verification procedures*: after data collection, techniques for verifying and comparing data need to be applied;
- *Perform data analyses*: this crucial stage involves the application of methodologies appropriate for answering key questions;
- *Conduct sensitivity tests*: to ensure robust results, there needs to be comprehensive testing for model specification, alternative inputs and outputs, and alternative methodologies;
- *Develop policy implications*: explore potential determinants of inefficiencies;
- *Engage in information dissemination*: reporting the results of performance comparisons helps engage stakeholders in the process;
- *Establish rewards and penalties*: procedures for planning and implementing incentives and corrective actions;
- *Seek public comments and promote stakeholder awareness*: such a campaign, with media assistance, promotes improvements in utility operations;
- *Prepare follow-up studies as part of the performance review*: benchmarking is an on-going process for monitoring and evaluating performance outcomes.

[9] This list and subsequent paragraph draws upon Berg and Holt, "Scorecards for Utilities and Regulators," *Water*, 21 April 2002, 51–53

5.2.2 Using performance scores

The water regulator can spearhead many of these activities. However, benchmarking will be much more effective if two criteria are met. First, regulated companies must cooperate with those responsible for providing oversight (whether the agency be a sector regulator or a municipal authority). Second, formal mechanisms need to be established for consumers and other affected parties to have access to rankings, raise concerns and suggest modifications to the process.[10] Through incentive regulation and/or an appropriate price cap formula, regulators can use findings from benchmarking reports to reward high performance companies. Studies that control for differences in physical, social and environmental contexts clearly show that average efficiency for water utilities improve with the "carrot" of incentive regulation. Incentives can motivate companies to improve efficiency, as they provide a vehicle to pressure laggard companies to promote cost-containment and the improvement of service quality. A properly designed benchmarking system should prevent poorly performing companies from increasing prices as much as the "average" water utility to which they have been compared—so long as prices, do indeed, cover costs. If companies operate more efficiently, customers will benefit from lower prices and should be able to expect and receive high quality service. The resulting system is likely to be sustainable—promoting further network expansion and the adoption of best practice by most water utilities. Regulators and managers who accomplish these tasks deserve high marks. Their legacy is a system in which their children and grandchildren will be justifiably proud.

5.2.3 Promoting public acceptance

The concluding point involves recognizing the physician's injunction to "Do no harm:" those conducting benchmarking studies need to avoid doing harm, damaging the credibility of performance comparisons. Stakeholder acceptance of the process requires equal doses of transparency, professionalism, and humility. It may be that quantitative analyses of cost functions and production functions are under-utilized by regulators and managers for fear that results will be misinterpreted by the public and the calculated efficiency scores will be misused by those monitoring performance. This concern can be addressed by seeking

[10] Benchmarking enhances transparency, but according to an analysis of the Netherlands, it impacted utility performance only after the data became publically available: Okke Braadbaart (2007). "Collaborative Benchmarking, Transparency and Performance: Evidence from the Netherlands Water Supply Industry." *Benchmarking: An International Journal*, **14**(6), 6770–6692.

more robust performance rankings, using techniques described in this volume. Reports should not claim excessive precision, but can certainly focus on the strongest and weakest performers. Other commentators are skeptical of simplistic (aggregative metric) performance comparisons. To the skeptics, one can say,

If not now, when?
If not here, where?

"It is likely that far more waste has occurred due to poor management practices (and weak incentives) in developing countries than to the misapplication of infrastructure benchmarking techniques."[11] The same point applies to water utilities in developed countries. Benchmarking is a fundamental requirement of good management. If managers do not have the data required for such comparisons, then one must question what they are actually managing. If regulators cannot identify historical trends, determine today's baseline performance, and quantify relative performance across utilities, then as an Indian regulator said, they may as well be writing "orders that are just pretty poetry."[12]

[11] C. Lin and S. Berg (2008). "Incorporating service quality into yardstick regulation: an application to the Peru water sector," *Review of Industrial Organization*, 32, 73.
[12] The quote is from an Indian regulator, cited in Bakovic, Tenenbaum, and Woolf (2003), "Regulation by Contract: A New Way to Privatize Electricity Distribution?" World Bank Working Paper, No. 14, p. 29).

Appendix 1 Variable Definitions and Explanations

Identifying variables and standardizing definitions has been a major activity within the global water community. Variable definitions are included here for completeness, with the IBNET *Indicator Variable Number* included in the description. The listing includes:

- Output Variables (with conversion factors).
- Quality Variables.
- Input Variables (quantities and prices).
- Conditioning/Environmental Variables (such as topology or population density).
- Macroeconomic Variables (per capita GDP and the unemployment rate).
- Governance/Structure Variables (such as incentive contracts and ownership).

This listing is very comprehensive, partly to illustrate the potential impacts of various factors on productivity or costs. A reasonable benchmarking study could not incorporate all these variables into the analysis. Start with a simple model and add complexity. If the analyst starts with a complicated model and lacks data, the study may not get off the ground!

Output Variables

Clearly, performance is partly based on outputs is delivered to customers. In addition, some studies consider the financial sustainability of the utility to be an output— for example, a high percentage of unpaid bills (for water delivered) would be considered a negative.

Output Variables for Water Services

Volume billed: Quantity of water annually billed to customers, measured in cubic feet, cubic meters, gallons, or liters per year. Since there are water losses due to leakages, evaporation, and robbing, the amount of water delivered is usually larger than the amount of water billed. This is the most intuitive measure of output, since it is the actual output that the firm sells and the process by which it gets its main revenues (bills not paid are also included in this variable). It can be split into residential and nonresidential (59 in IBNET).

Volume collected: This variable accounts only for bills actually paid by customers, and is measured in cubic feet, cubic meters, gallons, or liters per year. It can also be split into residential and nonresidential.

Volume delivered: Quantity of water annually delivered *into* the network, and is measured in cubic feet, cubic meters, gallons, or liters per year (55 in IBNET).

Volume purified: Quantity of water purified annually, measured in cubic feet, cubic meters, gallons or liters per year. Purification may have different meanings depending on the country or legal definition, but it is intended to capture chlorination and other treatments used in making water potable. It can also be split into residential and nonresidential.

Residential connections: Number of residential water connections of the firm. A residential connection is usually an account, but it is also possible to have a connection with multiple residential accounts, such as in the case of an apartment complex. This variable could be disaggregated into other types of residential connections, such as single- or multi-family connections. This is a widely used output variable (41 in IBNET).

Industrial connections: Number of industrial water connections of the firm. The distinction between residential and industrial connections is relevant from an output point of view because of the differing customer service issues. More about this is explained in the conditional variables section, where we distinguish the different features entailed by the characteristics of demand regarding customer profile, i.e., residential or commercial (41 in IBNET).

Metered connections: Number of water connections (industrial and residential) with an installed meter. This is not always used as an output variable itself, but if we consider that water metered is a different service that the utility is providing (as opposed to not metered water), it can perfectly be used to reflect this multiplicity of output. It is measured in number of metered connections (53 in IBNET).

Customers: Number of people that are actually served by the firm. It is usually larger than the number of connections, since a connection implies a household which can have more than one person. However, the terms customer and connection are often used interchangeably since the number of customers is always a multiple of the number of connections. This is one of the most commonly used output variables in the water and sewerage industry (40 in IBNET).

Customer-continuity: A more sophisticated way of approaching the number of customers is a combination of customer and continuity, where the continuity of the supply is considered. In this way, the product of each customer times the number of hours per day that the service is available is computed and summed up for all the customers.

Towns served: Reflects the number of towns actually receiving water services from the firm. It is a poor measure of output if considered alone but may help better describe the production process of the utility, especially in cases where the firm serves more than one city/town (34 in IBNET).

Output Variables for Sewerage Services

Volume collected: Quantity of sewerage annually collected, and is measured in cubic feet, cubic meters, gallons, or liters. It can also be split into residential and commercial volume (81 in IBNET).

Volume primarily treated: Reflects the quantity of wastewater annually treated in order to remove a significant proportion of solid matter. It is measured in cubic feet, cubic meters, gallons, or liters. It can also be split into residential and non residential (81d in IBNET).

Volume secondarily treated: Reflects the quantity of wastewater annually treated biologically in order to remove contaminants. It is measured in cubic feet, cubic meters, gallons, or liters (81e in IBNET).

Connections: As in water, this is the number of active sewerage connections of the firm. It can also be disaggregated into residential and industrial connections for the same reasons that the breakdown is done for water services (71 in IBNET).

Customers: As in water, this refers to the number of people that actually receive sewerage services by the firm. The same considerations apply (70 in IBNET).

Towns served: This variable reflects the number of towns actually receiving sewerage services by the firm. The same comments as in water apply (35 in IBNET).

Output Variables for Both Services

Network length: One variable used as a measure of output in some studies is the total length of the water/sewerage network that the firm utilizes. It is measured in miles or kilometers. This is not a very intuitive output variable since it is most commonly seen as an input for reaching customers that consume water. However, it is included in many studies as an output measure, so it can be considered a second option. The yearly change in network length can be a possible output measure as well, involving a dynamic approach that reflects the increase in potential customers that could be incorporated into the network (54 and 74 in IBNET).

Variable Conversions

Because different countries or regions sometimes measure the same variables in different units, it is useful to have a quick access to the equivalence between the most commonly used variables in the water and sewerage industry. Following here are the rules on how to go from one unit to another, focusing on the most commonly used variables.

Volume variables: Volume variables are used to quantify how much water is produced, delivered into the network, lost in the production-delivery process, purified before delivery into the network, and billed to customers.

To transform gallons into liters, multiply by 3.785412.
To transform liters into gallons, divide by 3.785412.
For example, 10 million liters of water is equivalent to 2.64 million gallons of water.

To transform cubic meters into cubic feet, multiply by 35.314667.
To transform cubic feet into cubic meters, divide by 35.314667.
For example, 20 million cubic feet of water is equivalent to 566,337 cubic meters of water.

To transform cubic meters into gallons, multiply by 264.172052.
To transform gallons into cubic meters, divide by 264.172052.
For example, 30 cubic meters of water is equivalent to 7,925.10 gallons.

Length variables: Length variables are mainly used to measure the size of the network, and to reflect the depth where water is available and/or extracted.

To transform miles into kilometers, multiply by 1.609344.
To transform kilometers into miles, divide by 1.609344.
For example, 200 miles is equivalent to 321.87 kilometers.

To transform meters into yards, multiply by 1.093613.
To transform yards into meters, divide by 1.093613.
For example, 350 yards is equivalent to 320.04 meters.

Area variables: Area variables help determine how densely populated the utility operation area is, as well as describing how big it is. They can also be used to measure the size of pipes utilized in the network.

To transform square miles into square kilometers, multiply by 2.589988.
To transform square kilometers into square miles, divide by 2.589988.
For example, 1200 square miles is equivalent to 3,107.99 square kilometers.

To transform hectares into acres, multiply by 2.471054.
To transform acres into hectares, divide by 2.471054.
For example, 3000 acres is equivalent to 1214 hectares.

To transform square inches into square centimeters, multiply by 6.4516.
To transform square centimeters into square inches, divide by 6.4516.
For example, 15 square inches is equivalent to 96.77 square centimeters.

For more information on quick conversions of measurements, three web sites are provided here:

http://www.onlineconversion.com/
http://www.digitaldutch.com/unitconverter/
http://www.convert-me.com/en/

Quality Variables

Consumers care about drinking water quality (whether the water they drink is healthy for them), continuity (whether it is available twenty-four hours per day or only intermittently), and mean time to repair (whether line breaks are repaired quickly). Similarly, measures of customer satisfaction (such as complaints per 1000 connections per year) are indicators of consumer perceptions. Determining service quality is an important step in rating a utility's efficiency and reliability. Quality of service (viewed as an output) can be measured using the following variables:

Drinking water quality: According to the WHO (World Health Organization) Guidelines for Drinking Water Quality, safe drinking water can be expressed in terms of microbiological quality and chemical quality.

Microbiological quality: The indicator is equal to the proportion of the analyzed drinking-water samples that fail to comply with the microbiological quality standards. In order to calculate the indicator, one needs data about the number of non-compliant samples and the total number of samples taken from a defined spatial unit (a water supply zone or other regional entity defined for regulatory purposes over the specific year). The "microbiological quality percentage compliance" can be computed as ([Total samples – Failed samples]/ Total samples)∗100.

Chemical quality: The indicator is equal to the proportion of the analyzed drinking-water samples that fail to comply with the chemical quality standards. In order to calculate the indicators, the analyst needs data on the number of non-compliant samples and the total number of samples taken from a defined spatial unit. The "chemical quality percentage compliance" can be computed as ([Total samples − Failed samples]/Total samples)∗100 (15.4 in IBNET).

Wastewater secondary treatment or better: Proportion of collected sewage that receives at least secondary treatment, e.g., removing oxygen demand as well as solids, normally biological. This indicator reflects the amount of effluent that is treated before being discharged (17.3 in IBNET).

Reliability: Reliability is a very important concern to most users. Reliability can be expressed as the proportion of the functioning time during a particular time period (e.g., a year). Functioning time is the number of days in a month or year when the water supply and/or sanitation service is functioning.

Continuity: Continuity of supply is equally essential. In many developing countries, the continuity of supply cannot be guaranteed. Of course, once most customers have adjusted to having service only a few hours per day (by making investments in storage tanks), the incremental benefits to improving continuity for current customers is reduced. Nevertheless, intermittent service can be viewed as an indicator of low quality of service. The continuity of supply is the average number of hours per day that people can get access to adequate amounts of safe drinking water in the home (15.1 in IBNET).

Non revenue water ratio: The distribution of water will invariably incur losses. The non-revenue water ratio is equal to the difference between water supplied and water sold as a percentage of water supplied. As Tynan and Kingdom (2002) point out, the unaccounted-for water ratio captures not only physical losses but also commercial losses, due to inefficient billing or illegal connections.[1]

[1] Nicola Tynan and William Kingdom (2002). "A Water Scorecard: Setting Performance Targets for Water Utilities," World Bank, Public Policy for the Private Sector, Note 242, April.

Thus, high levels of unaccounted-for water indicate poor system management and poor commercial practice as well as inadequate pipeline maintenance. Coelli *et al.* (2002) regard the water loss as an indicator of technical quality of service, which has been ignored by many studies (6.1 in IBNET).

Pipe breaks ratio: Total number of pipe breaks per year expressed as per km of water distribution network. This indicator measures the stability of the network system. A low pipe breaks ratio indicates good management and maintenance of the network (9.1 in IBNET).

Sewerage system blockage ratio: Total number of blockages per year expressed in per km of sewer. This indicator can reflect a number of issues such as the effectiveness of routine operations and maintenance activities, the hydraulic performance of the network, and the general condition of the pipes (10.1 in IBNET).

Mean time to repair: Mean time to repair is equal to the total time spent on repairs divided by the number of repairs in a given period. A smaler value indicates more prompt and efficient customer service.

Customer satisfaction: A more direct measure of service quality can be drawn from a customer survey. The satisfaction rate can be viewed as a good proxy of service quality. Another proxy variable is the complaint ratio which is equal to total number of customer complaints during the year divided by the number of customers (37a in IBNET).

One problem with some of the quality measures, such as reliability and the pipe breaks ratio, is that they may be influenced substantially by environmental shocks such as hurricanes and earthquakes, which may make the reliability figures of the utility companies in particular areas look quite bad. Therefore, regulators need to take these issues into account when they use the relevant indicators to conduct benchmarking studies.

Input Variables: Quantities and Prices

Input variables refer to internal factors (purchased inputs) that contribute to the overall effectiveness of the utility. They generally reflect decisions made by managers regarding resource use. These variables are as important as output and quality variables.

Labor—full time equivalent staff numbers: Total number of staff working at the utility on water and waste water service. Data are collected on the basis of full-time equivalent staff numbers because it can be problematic to ignore part-time employees. Using the full-time equivalent staff numbers imposes an implicit assumption that the skill intensity (human capital structure) is similar across firms and countries. However, different types of employees such as engineers, management, and skilled and unskilled workers have varying levels of ability, expertise, and experiences (human capital) (36 in IBNET).

Labor cost: The sum of salary and bonuses paid to employees. Some studies use the total salary expense to attempt to account for the differences in skill intensity (human capital structure) across firms. However, this method will be problematic if the relative wage rates for the various categories do not reflect their relative marginal product. Furthermore, this measure generates a new implicit assumption: the wage rates of different categories are similar across firms. This is a very strong assumption for international benchmarking (96 in IBNET).

Total number of staff-water: Total number of staff (in terms of full-time equivalent staff) working at the utility on water service. If we focus on the performance benchmarking of water supply, this leads to the use of the total number of water staff to measure the labor input (36a in IBNET). Of course, for water/wastewater utilities, allocating staff across the two activities can be quite difficult. Given the importance of understanding the extent of economies of joint production, particular attention needs to be given to the issue to avoid arbitrary allocations.

Total number of staff-wastewater: Total number of staff (in terms of full-time equivalent staff) working at the utility on wastewater service. Focusing on the performance benchmarking of wastewater service leads to the use of the total number of wastewater staff to measure the labor input (36b in IBNET).

Subcategories: A way around the potential problem due to the implicit assumption of uniform human capital structure is to use the subcategories of labor in the analysis. Labor can be further classified into subcategories such as management, engineers, other professionals, and skilled/unskilled workers. There are potential disadvantages to using this detailed classification. First, using the detailed classification in statistical analyses may cause the loss of several degrees of freedom, which is particularly important when the sample size is relatively small. Second, labor classification criteria may be different across firms or countries. Third, in many countries, this kind of detailed information is not available. It is probably better to have an approximation of the right number than a precise calculation of an inappropriate variable.

Labor price: Economic theory indicates the total cost of the firm depends on the output level and input prices. Therefore, we need to know the labor prices if we decide to use a cost function to compare the relative cost efficiency across firms. Labor price can be measured by the average wage of specific group of people.

For example, if we are using the total number of full time employment as labor input, the labor price can be measured by the average wage of full time employees, which is equal to the total salary paid to the full time employees divided by the number of full time employees. Similarly, if we have the data about executive numbers and compensation and we want to measure the labor price of management, we can calculate the average executive compensation, which is the total executive compensation divided by the number of executives (96 and 36 in IBNET).

Capital: Capital can be represented by physical or monetary measures. Both have strengths and weaknesses. Analysts should select the most appropriate measures based on country and industry characteristics.

Length of water distribution network: This measure is frequently used as a capital proxy. However, network length cannot capture all the capital elements used in the production process. In addition, it does not take into account other aspects of the network such as sizes (diameters) and material compositions. All of these are important factors that may influence the firm's cost and output (54 in IBNET).

Length of the sewer system: As above, when conducting a benchmarking study in wastewater service, length of sewer system can be used as a capital proxy. However, the same problems mentioned in the length of water distribution network apply (74 in IBNET).

Gross fixed assets: Gross book value of fixed assets at year end, including work in progress. Monetary measures may better reflect the capital used in the production process. A commonly used monetary measure of capital stock is the firm's fixed assets, which can be found in its financial statements. However, there are also limitations with monetary measures. It is difficult to get a consistent measure of capital due to the potential differences in accounting reporting methods, which is one of the major concerns in international benchmarking. In addition, many utilities in emerging markets have poor records or lack an asset registry. Therefore, most cross-country empirical studies use only physical inputs and outputs (112 in IBNET).

Gross fixed assets-water: The total gross fixed assets can be decomposed into gross fixed assets-water and gross fixed assets-wastewater. The gross fixed assets-water variable can be used to measure the capital stock (112a in IBNET).

Gross fixed assets-wastewater: Similarly, when conducting a wastewater service benchmarking study, the gross fixed assets-wastewater can be used to measure the capital stock (112b in IBNET).

Capital price: The calculation of capital price is also controversial. A simple version is to calculate the capital price as annual capital costs divided by capital stock. The capital cost can be measured as the sum of capital replacement (CAPEX), depreciation, and rental outlays for capital. The capital stock, as we illustrated above, can be measured in physical (network length) or monetary (net fixed asset) ways. Some researchers propose that we need to take the financing cost into account, which is the WACC (weighted average cost of capital) of the firm. WACC can be expressed as

$$WACC = \frac{D}{D+E}k_d(1-T) + \frac{E}{D+E}k_e$$

where D denotes debt, E is equity, k_d is the required rate of return for debt (interest rate), and k_e is the market-based required rate of return for equity, which can be calculated using the CAPM (Capital Asset Pricing Model).

Energy: Energy can be represented by physical or monetary measures. Regarding the physical measure, we can use the quantity of energy used, measured in thousands of kilowatts per hour (Mkwh). With reference to the monetary measure, the most commonly used monetary measure of energy is the firm's energy expenditure, which can be easily found in the income statements (97 in IBNET).

Energy price: The input price of energy input can be measured by the average price per kilowatt of electricity power in a specific year.

Chemicals: The most-used monetary measure of chemicals is the firm's chemical expenditure, which can be found in a detailed Income Statement.

Chemical price: The chemical price is equal to the chemical expenditure divided by the amount of chemicals used in the treatment process. However, the chemical price is often difficult to measure because the firm may have different chemicals in the treatment process. In these cases, the amount of water delivered is used as a proxy of the amount of chemicals. Therefore, the chemical price can be defined as the chemical expenditure divided by the amount of water delivered or treated.

Water loss: Water loss is considered an input by some analysts. It is the difference between the water delivered to the network and water sold to final customers. Alternatively, water loss can be viewed as an indicator of the effectiveness of past maintenance outlays and of the design and operation of the network. Given their outputs, firms will balance reducing water loss (input) to reduce pumping costs against the cost of identifying and reducing leaks. The variable for water loss is available from most utilities. (55 and 59 in IBNET).

Accounting/Financial Variables

Customer Payments and other sources of revenue are not usually viewed as inputs, except from the standpoint of financial sustainability. However, using these variables provides a glimpse into the firm's ability to efficiently maintain its level of services.

Revenues: This is the amount of money that the utility annually collects from the delivery and sale of water and sewerage services. It is measured in nominal, monetary terms. Revenue can include fixed fees charged to customers to have access to the network, and variable charges based on the volume delivered (for metered customers). The relative amounts will usually depend on the regulatory context and the utility's rate design, including up-front hook-up fees (or annualized fees). It should exclude any extraordinary cash flows from outside sources, like subsidies from international institutions or local governments. It can be split into revenues from residential and commercial customers (90, 90a and 90b in IBNET) or water and sewerage services. (90c and 90d in IBNET).

Soft loans and subsidies: Information on such cash flows would be very useful for analyzing the long term financial sustainability of a utility's operations. If such funds are off-book, the analyst will have to re-create the sources of external funds over time. Customers who are currently receiving service often have been subsidized in the past through such arrangements.

Total cost: This is the total annual cost of financing and operating the firm; from the standpoint of accounting, it includes operating costs, depreciation charges, and interest and tax payments. It is another monetary variable that draws upon the income statement of the firm. However, accounting statements may not include all the economic costs. Costs are usually disaggregated into their main components to provide a better idea of the scale of each element.

Operating costs: This item refers only to the cost of operating the utility in the provision of water and sewerage services to its customers. It does not include depreciation of fixed assets and interest payments to creditors or taxes to the government. Rather, it refers to the cost of providing services, regardless of how the activity is financed (i.e., what is the source of that money). It is mainly composed of salaries paid, energy consumed, administrative costs incurred, and maintenance of the network performed (excluding any investment toward a longer extension or a higher capacity). It is measured in monetary terms per period of operation, usually a year (94 in IBNET). It can also be separated into the two services provided: water and wastewater. (94a and 94b in IBNET.)

Labor costs: This variable refers to the salaries and benefits annually paid by the firm to its employees. It is part of the operating cost of the utility and is routinely the largest component (96 in IBNET). The treatment of set-asides for pension obligations may not be transparent. In some cases, pensions become unfunded obligations that represent future financial claims on company cash flows.

Energy costs: This variable reflects expenses on energy utilized by the firm to perform its operations. It is also part of the operating cost of the utility (97 in IBNET).

Depreciation costs: This is an accounting charge that is meant to reflect the annual loss of the value of fixed assets that have a long life of operation (usually decades). Since the amount annually assigned to depreciation charges is determined by accounting rules that may depend on the country where the utility is located, depreciation charges are sometimes excluded from cross-country studies. However, this number can serve as an indicator of the annualized capital expenses of the firm, or average age of fixed assets (when combined with balance sheet data). It is measured in monetary terms per period.

Taxes: This is the annual expense of the firm paid to the government in the form of income taxes. Income taxes are usually a fixed percentage applied to the net profit after excluding interest payments to creditors. It is important to note that these taxes do not include all those paid in order to provide water and sewerage services, like property taxes on the facilities the utility owns. Taxation will be quite different across countries and different for privately-owned and publicly-owned utilities within a single nation. Thus, publicly-owned firms may not always pay income taxes to the government, distorting the comparison with a privately-owned firm that is required to account for outgoing cash flow. Therefore, taxes raise important issues in inter-country (and inter-firm) comparisons.

Cash flow from operations: It is important to distinguish between cash flows resulting from operating the utility versus those cash flows associated with financing the utility. Apart from the net income annually generated by the utility from its standard operations, cash can also be added through the reduction of working capital. This will increase the amount of cash available for daily operations, and it is a variable that can also be used to perform financial analysis of the firm, usually dealing with its liquidity. It is measured in monetary terms per period.

Conditioning/Environmental Variables

Performance comparisons require that firms be "comparable" or that external variables be controlled for in the analysis. Conditioning and Environmental variables attempt to capture external factors that can affect utility operations. Some of these conditions exist outside the control managers; in other cases, managers can adapt to these external constraints.

General conditioning variables: Note that no model is likely to have many conditioning variables. The analyst will have to exercise some judgment regarding the relevance of unique factors that affect costs. However, the variables below illustrate the types of general conditioning variables that could be relevant in particular circumstances.

Range of service provided: These variables describe the types of services in which the firm is engaged, such as water supply and sewerage collection services. The firm could potentially provide other services as well (e.g., electricity and natural gas), and that should be reflected by these variables because economies of scope could be present due to the multiplicity of services. Economists define economies of scope as the ability of a firm to produce two goods or services at a lower average cost than two separate firms producing the same two goods or services. This may be the case for a firm engaged in both water provision and sewerage collection. Dummy variables are almost always utilized to account for this fact (32a in IBNET).

Resident population: This variable reflects the number of people living in the area served by the firm. The population may or may not be hooked up to the network or receive service. The variable indicates the potential customers the utility may have. It is measured in number of people. Depending on the type of regulation the utility faces, it may be required to meet a certain coverage percentage. Thus, the number of people living in the area has an effect on the minimum number of customers the firm is obliged to serve. From a dynamic point of view, the growth rate of the resident population may be an important variable to consider for the same reasons. Separation into water and sewerage is possible if necessary (30 and 30a in IBNET).

Population density: This variable accounts for how densely populated the area is under the firm's jurisdiction. It could reflect the possibility of the firm serving many customers with a shorter network where density of population is high. It is measured in number of people per square kilometer or square mile. Since these data are not always available, a proxy variable commonly used is the number of customers per mile of network. However, a problem with this variable is that the length of the network is not determined exogenously. Rather, it can be a choice variable for management if coverage targets are not mandatory.

Region of operation: The area where the utility is located could affect availability of water. Transportation and pumping costs are not the same if the utility is located next to a river (Buenos Aires, Argentina) or far away in the mountains. This fact should be taken into consideration in comparing costs (or production) across a set of firms. Such a conditioning variable is usually incorporated into models through dummy variables, so there is no natural unit of measurement.

Peaks and lows of demand: If demand differs with each season (e.g., summer vs. winter) or at different times of the day, this can have a significant impact on the design of the whole network and its operating conditions. This seasonal volatility of demand should be taken into account, since the network has to be able to handle a variety of demand conditions. Such conditions can imply a final network size that operates at full capacity during the high periods of demand, but not fully utilized during low-demand periods. Averaging monthly production is not a good proxy for output in this case. Furthermore, it may be important to capture this feature of the external environment. It could be measured as a ratio of high/low demand, measuring demand in volume units (liters, gallons, or cubic meters).

Ratio residential vs. commercial customers: The proportion of residential customers to commercial customers characterizes the type of demand faced by the utility. A higher proportion of residential customers generally results in lower consumption per customer but a greater amount of customer service, meter reading, and billing activity by the firm. A higher proportion of industrial customers, on the other hand, results in much higher consumption for each connection, with reduced meter-reading and billing activity. The quality of the water can also differ across types of customer served; and that could be reflected in this variable. This variable can be measured in many ways, like percentage of residential customers or ratio of number of residential customers to number of commercial customers, among others.

Weather and Topographical Variables

Rainfall: The amount of rainfall in the area where the utility is located could affect the availability of water and is accounted for with rainfall variables (average, maximum or minimum; for periods of, for example, the last 12 months, 24 months, or 5 years). A rainy area can make underground water more easily available, as well as water taken from a river or lake. Rainfall is measured in inches or millimeters per year.

Temperature: Temperature is another weather condition that affects the performance of water utilities and could be taken into account in a benchmarking study. It is measured in Fahrenheit or Celsius degrees. It can refer to the last year or the last 5 years (the same as rainfall), and it can reflect the average, the maximum, or the minimum temperature in the area served by the firm. One of the effects of a higher temperature is more evaporation in the water deposits of the firm, but it can also affect consumption by customers.

Topology: This variable reflects the difference between the highest and the lowest areas served by the firm (with respect to sea level), and it can be included to reflect the difficulty for the firm to pump water up and down within the jurisdiction where it is operating. It is measured in meters or yards. This variable is not widely used but it can be helpful.

Macroeconomic Variables

For some types of studies, other exogenous variables can be taken into account—especially if observations are available for longer time periods.

Per capita GDP: This variable can reflect different propensities to spend money on water by the population served. Ability to pay makes a difference to the cash flows that might be obtained by a firm. A higher income could affect water consumption (and willingness to pay) in a variety of ways, and this variable takes that fact into consideration. For example, residential consumption by high-income people can be higher due to the existence of washing machines that low-income families cannot afford. It is measured in dollars per year per person, or any equivalent measure. It is a variable that could be important in a long term, dynamic study but is not in a short-term analysis, especially when dealing with a small area where people are likely to have similar incomes (50 in IBNET).

Unemployment rate: The unemployment rate is a different way to measure consumption characteristics of the people served by the utility. Presumably, a higher unemployment rate in the area under consideration can have a negative impact on consumption. It is measured in percentage terms. It has the same interpretation as the per capita GDP.

Governance Structure Variables

These variables might be used by academic researchers to examine the impacts of particular governance systems, ownership, and regulatory incentives. They are listed here to illustrate the range of issues that might be investigated.

However, studies of relative performance would not include such variables, since a low performing firm (with poor internal incentives) might appear to be high performing relative to other firms once since the poor incentives were taken into account. The point of benchmarking is to identify actual performance. Explaining that performance in terms of governance structure then becomes another step in the policy analysis process.

Performance (incentive) contracts: The performance contracts between government and utility managers are important sources of internal incentives. Management could be motivated to improve performance through bonuses based on the achievement of targets, while poor performance could be penalized through firing or demotion. Specifically, contracts can improve performance and efficiency if they: (1) reduce a manager's information advantage; (2) provide sufficient incentives for managers to increase effort; or (3) manifest the government's commitment to honor its promises. Despite its importance, very few studies have explored the impact of performance contracts on water utilities' performance. Studies of the water system in Uganda find that performance contracts have improved efficiency and productivity growth.

Regulatory mechanisms: Different regulatory methods may influence firms' incentives and, hence, their efforts and outputs. Using the data from telecommunication and electricity industries, many studies show that different regulatory mechanisms such as price cap and rate of return structures have quite different impacts on firm efficiency, service quality, and network modernization. Regarding the water industry, to our knowledge, one study explores the relationship between regulatory mechanisms and firm efficiency using the data (from Wisconsin water utilities). The results show that different regulatory mechanisms exert significant impacts on firm efficiency.[2]

Rate of return: A dummy variable which takes a value of 1 if a firm is under rate of return regulation. Rate of return regulation assures firms get a "fair" level of return on capital. Regarding rate of return regulation, the direct link between changes in costs and changes in prices (after a rate review) insulates the firm from adverse shocks, but also weakens the firm's incentive to pursue cost containment (R1 in IBNET).

Price cap: A dummy variable which takes value 1 if a firm is under price cap regulation. The regulator sets a predetermined maximum price or maximum

[2] Steven Renzetti and Diane P. Dupont (2009). "Measuring the technical efficiency of municipal water suppliers: the role of environmental factors," *Land Economics*, **85**, 627–636 (November).

price increase. Therefore, any cost savings obtained due to higher effort and efficiency improvement accrue to the firm. Under the price cap regulation, utilities have strong incentives to devote effort towards cost-minimization efforts, improving efficiency. However, under price caps, the incentive to improve quality is dampened unless quality standards are mandated (and penalties applied for not meeting targets (R1 in IBNET).

Other regulations: Other regulatory mechanisms include revenue caps, revenue sharing regulation, and hybrids of rate of return and incentive regulations (R1 in IBNET).

Corruption: In cross country studies, corruption has been shown to negatively affect water utility performance. Transparency decreases the likelihood of corruption, so benchmarking can be one instrument for limiting opportunities for bribery and other forms of corruption. Corruption is also associated with political instability. A cross-section analysis of 85 nations (1990–2004) found reduced access to improved drinking water and sanitation for nations with higher indices of corruption.[3] Along the same lines, Estache and Kouassi (2002) conclude that if African water utilities operated in a corruption free environment, production costs would fall by 64%. Thus, this variable (represented by the Kaufmann Index or one or more of the Worldwide Governance Indicators) has been used in academic studies of infrastructure performance. Another index is available from the International Country Risk Guide (ICRG), published by Political Risk Services Group.

Ownership Identity: Privatization has become a global trend in the past 25 years. The cumulative value of funds raised through privatization in more than 100 countries exceeds 1.5 trillion dollars. Much research has been conducted that compares pre- and post-privatization performance changes or the relative performance between state-owned and private firms. Overall, the evidence supports the proposition that private ownership is associated with better firm performance than state ownership because state ownership is often associated with a series of problems, such as weak incentives, inadequate monitoring, soft budget constraints, and political or personal objectives other than cost-minimization (linked to profit maximization). In the water sector, several studies have shown mixed results. Therefore, ownership structure is often incorporated into benchmarking studies to see the differential impact of

[3] Nejat Anbarci, Monica Escaleras, and Charles A. Register (2009). "The ill effects of public sector corruption in the water and sanitation sector," *Land Economics*, **85**, 363–377 (May).

ownership. The following variables can be used to describe a firm's ownership structure.

State ownership: This is the percentage of the firm's total shares (or total assets) held by the state government in a specific year (3a in IBNET).

Private ownership: This is the percentage of the firm's total shares (or total assets) held by private investors in a specific year (3a in IBNET).

Foreign ownership: This is the percentage of the firm's total shares (or total assets) held by foreign investors in a specific year. For utility companies in developing countries, foreign ownership is expected to exert a positive impact on the firm's performance for the following reasons. The introduction of foreign investors might result in better monitoring of management. Furthermore, foreign investors may introduce advanced technology and managerial skills into the utility companies. On the other hand, local unrest with price increases can lead to outbreaks of civil disobedience directed at international firms.

Ownership concentration: Besides the ownership identity variables listed above, ownership concentration may have important impacts on the firm's efficiency, especially in developed countries where many utility companies are publicly-listed. According to some previous studies, a majority shareholder can have two very different effects on firm performance. The first effect arises because, compared to small shareholders, a major shareholder has stronger incentives and more power to monitor management and improve firm performance. On the other hand, the major shareholder might collude with the managers to expropriate the minority shareholders' benefits or capture other private benefits of control at the expense of smaller shareholders. Two variables are frequently used to describe the ownership concentration: controlling shareholder concentration and summed ownership concentration.

Controlling shareholder concentration: The ownership percentage of the largest shareholder of the firm.

Summed ownership concentration: Total holdings of all shareholders whose ownership percentage is greater than 5%.

Board structure: The Board of Directors represents an important internal corporate governance mechanism to help resolve agency problems that arise from the separation of ownership and control in modern corporations and improve firm performance. Three variables are frequently used to describe the board structure: board size, number of outside directors, and CEO responsibilities.

Board size: Total number of directors on the board. Some researchers argue that larger boards are less effective in monitoring management and are more susceptible to the influence of the CEO. However, this is not a general conclusion for all cases and evidence is mixed.

Number of the independent (outside) directors: Independent directors are believed to have an incentive to monitor management and ensure the effective running of the company since being a director of a well-run company builds their reputations as experts in decision control and signals their competency to the market.

CEO duality: CEO duality is a dummy variable, which takes value 1 if CEO is also the chair of the board of directors. Agency theory suggests that holding the board chair position enables CEOs to exert greater influence over board agendas and decisions, which weakens the monitoring function of the board.

Compensation structure: Compensation structure is an important internal incentive mechanism to help resolve agency problems that arise from the separation of ownership from operational control. The compensation structure can stimulate management to perform better. The performance contract mentioned above is an important aspect of executive compensation.

Stock ownership and stock option: Stock-based incentive compensation is widely adopted in developed countries. Stock option and stock ownership tie the management's income to firm performances and align the management's interest to that of the shareholders. Stock-based compensation provides long-term incentives to the management to improve firm performance. Regarding the stock ownership, percentage of outstanding shares owned by the executive team is a frequently used measure. With regard to stock option, the Black-Scholes option pricing model is frequently used to determine the current value of the option.

Pay gaps (pay dispersion): There are two views about pay dispersion within the firm. Tournament theory argues that the large pay dispersion among corporate hierarchies, combined with uncertainty over who will be promoted, provides managers with strong incentives to compete in the "tournament" and exert considerable efforts to win the prize. On the other hand, equity theory and sabotage theory argue that firms with less variance in compensation will have more harmonious and cohesive labor relations, and thus achieve higher efficiency. Furthermore, large pay dispersion might reduce team cooperation and even induce sabotage behavior. Two measures are frequently used to describe the level of pay dispersion: (1) pay gap between first tier and second tier managers and (2) Coefficient of Variance of the executive compensation

Pay gap between first tier and second tier managers: Pay differences between the first tier managers (e.g., CEO and board chair) and the second tier managers (e.g., Vice Presidents).

Coefficient of Variance of the executive compensation: The variance of executive compensation divided by the mean of the executive compensation.

Concluding Remarks about Variables

Ultimately, the question being asked and data availability determine which specification and which methodology are most appropriate. Thus, the definitions presented here (and associated *caveats*) are meant to illustrate the range of potential data needs for conducting benchmarking analyses or follow-up studies. IBNET, the International Water Association, and other groups have tried to design data collection systems that allow information to be gathered on key variables. None of the studies summarized in Appendix 2 (Bibliography) uses all the variables listed here.

Appendix 2 Annotated Bibliography of Water Benchmarking Studies

Each of the eight studies summarized below has features that are worthy of emulation. However, each study has its own limitations. They are briefly surveyed here to illustrate the types of studies being conducted by researchers. The summaries identify some researchers' choices when facing sample size (and other data problems) and the use of different methodologies. The implications of these studies can only be alluded to — the results are far too rich and complex to be captured in a short annotated bibliography. The journal articles and book chapters are meant to be illustrative; this is not a comprehensive catalogue of available cost and production function studies.[1]

Peru: Corton, Maria Luisa (2003). "Benchmarking in the Latin American water sector: the case of Peru," *Utilities Policy*, **11**, 133–142.

One of the early developing country benchmarking studies was possible because the Peruvian water regulator, SUNASS, had made data available on-line. Corton's study was the first of a series of benchmarking studies of water utilities in that country. Using data from 44 water utilities in 1998, this study evaluates their relative performance. A cost function is estimated using Ordinary Least Squares (OLS) and the residual is interpreted as comprised only by inefficiency — the simplest way to measure inefficiency. Operating cost is a function of volume of water produced, network size (length of mains). The author also examines number of districts administered by each company to explore its effect on operating costs. In addition geographical characteristics

[1] Murillo-Zamorano and R. Luis (2004). "Economic efficiency and frontier techniques," *Journal of Economic Surveys*, **18**(1), 33–77. This study provides a critical and detailed review of parametric and non-parametric frontier methods, not only technical efficiency but also cost efficiency measurement. The authors analyze non-parametric models such as DEA, and analyze parametric methods such as SFM; other topics such as Bayesian techniques, bootstrapping, and duality theory also are considered.

are controlled for by the introduction of dummy variables for regions (coast and mountains, with the intercept capturing the cost implications of being located in the forest region). The relative efficiency for each utility is derived from the model. Furthermore, the study provides a detailed description of the institutional background of Peru's water sector and the benchmarking methods adopted by the water regulator (SUNASS). The ranking based on the OLS model differs from the ranking based on the SUNASS evaluation system. However, the latter includes more factors than operating efficiency.

A second study extends her work using data for 44 water utilities during the period 1996 to 1998, evaluating the consistency of water utility performance rankings for Peruvian water utilities.[2] The results indicate that Data Envelopment Analysis (DEA) and Stochastic Frontier Analysis distance function models yield similar rankings in this case. In addition, the techniques have comparable success for identifying the best- and worst-performing utilities. Again, these rankings, based on sophisticated statistical techniques, are not highly correlated with those developed by the Peruvian water regulator (SUNASS). The study analyzes the potential reasons of the ranking difference, such as the inclusion of additional performance indicators, the equal-weight method used in the SUNASS evaluation system, and the lack of input-output causative relationships in the SUNASS system. The study illustrates the importance of developing sensitivity tests for identifying weak utilities. Another study (Lin, 2005) studies how the introduction of quality variables affects performance comparison across utilities. This study utilizes data from Peru for the period 1996–2001, applying stochastic cost frontier methods.[3] The author concludes that quality variables should be incorporated as outputs in a cost function, rather than as environmental variables.

Brazil: H. Tupper and M. Resende (2004). "Efficiency and regulatory issues in the Brazilian water and sewage sector: an empirical study," *Utilities Policy*, **12**, 29–40.

Brazil's data collection effort (SNIS) is part of the Modernization Program of the Sanitation Sector (PMSS), which Brazil started in 1992 with financial support from the World Bank. This initiative provided the information foundation for a number of studies. Tupper and Resende quantify the relative efficiency of 27 state water and sewerage companies in Brazil during the period 1996–2000. Relative

[2] S. Berg and C. Lin (2008). "Consistency in performance rankings: the Peruvian water sector," *Applied Economics*, **40**(4–6), 793–805.

[3] Lin, Chen (2005). "Service quality and prospects for benchmarking: evidence from Peru water sector," *Utilities Policy*, **13**, 230–239.

efficiency scores obtained using a DEA methodology suggest that some utilities have suboptimal performances. Controlling for regional heterogeneities, the complementarities between DEA and econometric procedures are explored by controlling for network density and water loss factors. The procedure leads to adjusted relative efficiency scores that are considered in the construction of an incentive mechanism: a linear reimbursement rule. The comparison of actual costs and reimbursed values defined by the comparative efficiency analysis indicates which companies have the greatest opportunity (and rewards) for cost savings. Later studies have used larger samples to examine relative performance of water and sanitation utilities. Seroa da Motta and Moreira (2006)[4] use DEA to analyze a sample of 104 utilities (1998–2002); they find that regional (state) operators benefit from larger scale economies than municipal operators. A later study by Sabbioni (2008) examined the relative efficiency of regional and three different types of local (municipal) enterprises in providing water and sewerage services in Brazil between 2000 and 2004.[5] Using an unbalanced panel data with 1200 firm-year observations over the time period 2000 to 2004, he estimates the ownership effects on efficiency using a two stage process based on a cost function with fixed effects. Perhaps the most important lesson from this series of studies is the value of a serious national data collection effort.

England and Wales: D. Saal and D. Parker (2005). "Assessing the performance of water operations in the English and Welsh water industry: a lesson in the implications of inappropriately assuming a common frontier," in T.J. Coelli and D. Lawrence (eds.), *Performance Measurement and Regulation of Network Utilities*, Edward Elgar, Publisher, pp. 297–327.

This study represents one of the authors' many contributions to the analysis of water utilities in England and Wales.[6] It illustrates the kinds of hypotheses tested by scholars and the application of an input-distance function in a panel

---

[4] R. Seroa da Motta and A. Moreira (2006). "Efficiency and regulation in the sanitation sector in Brazil," *Utilities Policy*, **14**, 185–195.

[5] Sabbioni, Guillermo (2008). "Efficiency in the Brazilian Sanitation Sector," *Utilities Policy*, **16**(1), 11–20.

[6] A sample of other studies would include D. Saal and D. Parker (2000). "The impact of privatization and regulation on the water and sewerage industry in England and Wales: a translog cost function approach." *Managerial and Decision Economics*, **21**(6), 253–268. D. Saal and D. Parker (2001). "Productivity and price performance in the privatized water and sewerage companies of England and Wales." *Journal of Regulatory Economics*, **20**(1), 61–90. D. Saal and D. Parker (2004). "The comparative impact of privatization and regulation on productivity growth in the english and welsh water and sewerage industry," 1985–1999. *International Journal of Regulation and Governance*, **4**(2), 139–170.

context. They decompose productivity growth to account for the impact of technical, efficiency, and scale changes on performance. They demonstrate that in that industry, the water only and the water/sewerage companies do not share a common frontier. They utilize a translog input distance function to derive a productivity index and its decomposition. Water supply and physical connections are taken to be the two outputs. Their measure of capital is the "modern equivalent asset" estimate of the replacement cost of net tangible physical assets—obtained from the annual regulatory accounts of the utilities. Other factors affecting costs (beyond managerial control) were included in their model: density (connections per kilometer of water mains), average pumping head (lift times the volume of water), and water quality (percentage of water in compliance with nine key water parameters). Including a few other variables resulted in estimating 20 parameters. Results from model estimates indicate that as density increases, input requirements fall and that as topographic conditions worsen, input requirements increase. However, the sign on the water quality variable is counter-intuitive, raising questions about the specification of this particular model. The authors find that for a separately estimated (water-only) frontier productivity increased by 20.5% between 1993 and 2003.

This study was intended more to illustrate the application of the distance function and a number of issues associated with model specification, in particular separability of the cost function. It does not attempt to derive X factors for individual utilities. The results provide a cautionary lesson for regulators who attempt to evaluate the performance of firms producing multiple outputs when data is limited for some of them. Saal and Parker conclude, "Thus we would suggest that there is a substantial tension between improving the tractability of regulatory models by focusing on more readily estimable function-level performance assessment, and the potential bias of such models because of their implicit assumption that firms are separable at the functional level." (p. 322)

Australia: Coelli, Tim and Shannon Walding (2006). "Performance measurement in the Australian water supply industry: a preliminary analysis," in T.J. Coelli and D. Lawrence (eds.), *Performance Measurement and Regulation of Network Utilities*, Edward Elgar, Publisher, pp. 29–55.

The main purpose of this study was to assist regulatory authorities in setting CPI-X price trajectories, focusing on the estimation of Total Factor Productivity performance for Australian regulatory authorities. The authors utilized annual data on 18 firms for 1995/96–2002/03 (from the Water Supply Association of Australia—WSAA) and calculated technical and scale efficiency scores for the last year. In their DEA analysis of the panel of firms, they specified a model with two output variables (connections and water delivered) and two input variables (operating expenditure—OPEX and capital). Capital is taken to be the depreciated capital stock, with total length of water mains tested as an alternative capital variable—for robustness. The latter proved to be a better measure of capital in this case. The mean technical efficiency score for the 18 firms was 0.904, implying that to obtain the same output levels, input usage could be reduced by 9.6 percent for the average firm. They also ran a second-stage regression, checking whether the technical efficiency scores could be explained by other factors, like percentage of non-residential connections, average rainfall or other variables. None were significant at the 10 percent level.

Using the Malmquist TFP (total factor productivity) index, Coelli and Walding estimate the TFP change, which can be further decomposed into technical change (frontier shift) and technical efficiency change (catch-up). They obtained a weighted mean annual total factor productivity growth of 1.1 percent. In their discussion of implications for price regulation, they provided illustrative calculations of X-factors, where firms with low efficiency scores had higher X factors, as they were required to catch up to the industry average score over the (assumed) five year price review time horizon. The authors are careful to point out that the numbers obtained for each of the 18 firms should be used as "starting points" for discussions between the service provider and the regulator. Unique circumstances could explain why a firm achieved a low efficiency score, which should be taken into account when establishing prices based on performance comparisons.

Italy: B. Antonioli and M. Filippini (2001). The use of a variable cost function in the regulation of the Italian water industry, *Utilities Policy*, **10**, 181–187.

Utilizing panel data for 32 provincial water distribution firms in Italy (1991 to 1995), the authors estimate a variable cost function that suggests the existence of economies of output and customer density, along with the presence of small diseconomies of scale. The paper incorporates some service area characteristics not taken into account by the regulatory authority, indicating ways to improve the actual tariff regulation process. The total variable cost equation utilizes output, price of labor, size of the water system (length of network), number of customers, a proxy for capital stock (number of water wells), water losses and a zero-one dummy variable capturing the need to treat source water to meet health standards. The last two variables are both expected to increase costs. Thus, here, number of customers is taken to be a factor affecting costs rather than as an output. The authors use a Cobb-Douglas function as opposed to a translog functional form to reduce the presence of multicolinearity in the model. The Italian regulatory authority uses a similar model for yardstick regulation but the results of this model show some differences since different variables are included in the model. The authors point out how differences in parameter estimates might occur when utilizing different variables. Consequently, the authors stress the importance of performing robustness checks after estimation, particularly when the results are to be used by a regulatory authority. Thus, these academic researchers provide some support for the utilization of benchmarking in the regulatory process and suggest ways to enhance current procedures.

Uganda: Mugisha, Silver (2007). "Performance assessment and monitoring of water infrastructure: an empirical case study of benchmarking in Uganda," *Water Policy*, **9**(5), 475–491.

One of Africa's success stories in water utility performance is documented by Mugisha in this analysis of initiatives taken by the National Water and Sewerage Corporation (NWSC), Uganda. The empirical portion of the study adopts a SFA production function model to estimate firm efficiencies, using maximum likelihood estimation techniques. A Translog model is utilized and is shown to be a better representation of the situation than the Cobb-Douglas model (which has fewer parameters). Because of data limitations during some time periods, the author uses an unbalanced panel (ranging from eight to fourteen utilities) for 1996–2004. Output is taken to be water billed; inputs include delivered water (so water losses will reduce output), number of connections, network length, and total staff. Market size (target population) and commercial/customer orientation were used as a control variable to capture environmental factors, where the latter represented a shift in managerial attention—from an engineering orientation to one that focused on customers and commercialization. The inclusion of market size as an environmental variable in the model allows the analyst to predict how companies would be ranked if they were operating in comparable environments. Using time as a variable, the author is able show that the rate of technical change has been positive—though there is still a need to disentangle movements towards the frontier from shifts in the frontier itself (which reflect technical change). While performance scores are not summarized, such information could be derived for individual utilities. The quantitative work has been used in conjunction with other studies to establish targets for managers in NWSC incentive compensation schemes.

Cross Country: C. Nauges and C. van den Berg (2008). "Economies of density, scale and scope in the water supply and sewerage sector: a study of four developing and transition economies." *Journal of Regulatory Economics*, **34**, 144–163.

Cross country studies raise issues of data comparability and differences in underlying governance structures and physical conditions. A careful study by

Nauges and van den Berg examined four countries at different stages of economic development and with very different institutional and geographic characteristics: Brazil, Moldova, Romania and Vietnam. The data were not pooled, but parameter estimates were obtained for each country. The focus of the research is on measures of density, scale and scope, using a Translog cost function. They find economies of scale for utilities and economies of density in the last three nations; in addition, they find economies of scope in Brazil (1996–2004), Moldova (1996–2004), and Romania (2000–2004), where the number of years is given in parentheses. Sample sizes are 26, 38, and 23 multi-product utilities, respectively. Data for 67 water-only utilities in Vietnam were available from 1997–2000, so these were single-output firms (volume of water delivered). Total variable cost is taken to be a function of output levels and input prices (such as labor and energy). Control variables included the length of water mains and the population connected to the water network. Depending on the data available for a country, other variables were included in the models: for example, percent metered and pipe breaks are used as indicators of service quality. Note that other studies have often taken the number of customers connected to be a separate output (rather than a control variable). Quality of water supplied was not included in the models, due to lack of such information in the IBNET dataset (used in this study). The authors recommend that future studies utilize such cost studies to examine the impacts of different regulatory governance and incentive schemes.

Cross-Country in a Region (Asia): A. Estache and M. Rossi (2002). How different is the efficiency of public and private water companies in Asia? *The World Bank Economic Review*, **16**(1), 139–148.

Using 1995 data from a sample of 50 water companies in 29 Asian and Pacific region countries, this study provides further evidence on the differences of efficiency between public and private utilities. A stochastic cost frontier Cobb-Douglas approach is used. Operational costs are taken to be a function of number of clients, daily production, and number of connections (output variables), number of hours of water availability per day (quality measure), average salary, population density, surface source as a percent of total water, percentage metered, and a few other control variables. The study represents an early attempt to capture the roles of different factors on costs. The attention given to performance by the twenty-two private sector utilities and the twenty-eight public sector utilities reflects a policy concern regarding relative efficiency. The results indicate that efficiency is not significantly different between private and public utilities for this sample during this time frame. However, the authors emphasize the need for data collection efforts.

The study also reviews the methods of introducing benchmarking into the regulatory process. The results stress the advantages of relying on efficiency frontiers over the traditional partial indicators because the frontier methods can consider both multiple inputs and outputs simultaneously and can control for factors over which the operators have no control (e.g., diversity of water sources or of raw water quality, characteristics of users). For instance, in the Working Paper on which the published study is based, the authors find that Dhaka is ranked first when using the stochastic frontier model, and forty-third when using a more traditional indicator, such as accounts receivable or numbers of workers per 1000 connections.

Although a wide range of studies are becoming available to decision-makers, they have had a limited impact actual rate cases or incentive schemes—except in a few countries. Thus, there are significant opportunities for additional benchmarking studies that will shed light on water utility performance patterns in individual countries and in regions of the world; any serious initiative will need to be accompanied by comprehensive data collection efforts.

Appendix 3 Technical Features of Benchmarking Methodologies

Partial Indicators (Specific Core Indicators)

Partial indicators (PIs) can provide a useful starting point for identifying poor performers. Such "core indicators" include accounts receivable rates, cost recovery rates, water loss rates, or volume delivered per worker; these ratios can be used to capture specific aspects of utilities' operating and financial performance. Simple trend analysis can draw a rough picture of firm performance change over time. As has been noted, such indicators are sometimes combined to create an Overall Performance Index. As an intermediate approach, partial indicators can also be combined to create an index variable for a statistical model. For example, in a study of Austrian utilities, several partial indicators were combined in order to determine "urbanity," which constitutes one variable within the benchmarking model developed by researchers.[1]

Data Requirements: Relevant variables used in the calculation of the ratios.
Advantages: PIs are easy to calculate; results are intuitive; the analyst can conduct a comparison with only two observations.
Disadvantages: PIs only reflect one aspect of firm's operating or financial activities; the comparisons do not control for the impacts of other factors. When they are aggregated into an Overall Performance Index, the weights applied to each component are often not based on explicit policy priorities.

[1] H. Theuretzbacher-Fritz, R. Neunteufel, J. Kolbl, R. Perfler, and E. Mayr "The right choice of denominators and grouping factors in water supply metric benchmarking," in Cabrera and Pardo (2008), *op. cit.*

Total Factor Productivity (TFP) Index

TFP indices focus on the productivity change over time. The indicator attempts to account for increases in output that accrue due to technological change and efficiency improvements, rather than due to increases in inputs (where the quality of those inputs could also affect output). In its simplest form, it is just outputs divided by inputs. Changes in this ratio over time reflect changes in the productivity of those inputs. We will not go into the precise calculations of these indices here; the two commonly used TFP indicators are the Tornqvist and Malmquist indices.[2]

Tornqvist index

Data requirements: quantity and price data on inputs and outputs for two or more firms or time periods.
Advantages: not difficult to calculate; prices are used as weights; captures allocative efficiency; analyst can conduct a study with only two observations.
Disadvantages: need price information for inputs and outputs; TFP change cannot be further decomposed into technical efficiency change, technology change, and scale efficiency change.

Malmquist index

Data requirements: quantity data on inputs and outputs for firms in different time periods.
Advantages: price information not needed; based on DEA model, so no functional form assumption; TFP change can be decomposed into technical efficiency change, technology change, and scale efficiency change.
Disadvantages: needs relatively large sample size for each time period; the panel has to be balanced (same firms appear in every period).

Non-parametric Methods

The most commonly used non-parametric methods are Data Envelopment Analysis (DEA) and related models. As noted earlier, DEA is a linear

[2] An accessible overview of related topics is available in a book by Tim Coelli, Antonio Estache, Sergio Perelman, and Lourdes Trujillo (2003). *A Primer on Efficiency Measurement for Utilities and Transport Regulators*, World Bank Institute xii–134. For a more technical volume, see Tim Coelli, D. S. Prasada Rao, Christopher J. O'Donnell, and George E. Battese (2005). *An Introduction to Efficiency and Productivity Analysis*, Second Edition, Springer Publishing, xvii–349.

programming method that constructs a nonparametric production frontier by fitting a piece-wise linear surface over the data points. The technical-efficiency based DEA models can be constant returns to scale (CRS) and/or and variable returns to scale (VRS) models. Other DEA models include DEA with preference structure, DEA with undesirable outputs and DEA with environmental variables.

DEA production returns to scale models: DEA models can be specified to be Constant or Variable Returns to Scale.

> *Data requirements*: quantity data on inputs and outputs for the sample of cross-sectional firms or the panel. The panel sample does not need to be balanced.
> *Advantages*: no functional form assumption; can easily handle multiple inputs and outputs.
> *Disadvantages*: requires relatively large sample size; attributes all the deviations to inefficiency (no random noise); as the complexity (number of variables) in a DEA model increases, the discrimination power of DEA models decreases; more firms tend to be on the efficiency frontier.

Cost efficiency DEA model: If price information of the inputs is available, the analyst can conduct a cost efficiency DEA model to calculate the cost efficiency of the firms. Since cost efficiency is the product of technical efficiency and allocative efficiency, the analyst can calculate firm-specific allocative efficiency based on the cost efficiency DEA model and regular VRS/CRS DEA models.

> *Data requirements*: quantity data on inputs and outputs and price data on inputs for cross-sectional firms or panel sample; panel sample does not need to be balanced.
> *Advantages*: can be decomposed into allocative efficiency, technical efficiency, and scale efficiency.
> *Disadvantages*: same as those listed for CRS/VRS DEA models; it is often difficult to obtain input prices.

Super efficiency DEA model: Results may yield multiple efficient utilities (efficiency score = 1) in the CRS/VRS models. This outcome is especially likely if the sample size is small relative to the number of variables used in the DEA models. The super efficiency model is one method used to rank the efficient utilities. The basic idea of super efficiency model is to compare the efficient firm to the frontier constructed from *all other* firms (excluding the firm itself) and evaluate the relative efficiency superiority of the firm to the others.

> *Data requirements*: quantity data on inputs and outputs for cross-sectional firms or panel sample; panel sample does not need to be balanced.

Advantages: can rank the efficient utilities in CRS/VRS models.
Disadvantages: the firms are not compared to the same "standards" because the frontier constructed from the remaining firms changes for each efficient firm being evaluated.

Parametric Methods

Ordinary Least Squares is the simplest regression technique yielding estimates of key parameters in production or cost functions. In academia, the most commonly used parametric methods are Stochastic Frontier Analysis (SFA) and corrected ordinary least squares (COLS) models. The main difference between these models is that COLS attributes all the deviations from the frontier to inefficiency while SFA models attribute *part* of the deviations to inefficiency and part of the deviations to random noise. In other words, the SFA models take both inefficiency and random noise into account. The most widely used stochastic frontier models include the stochastic production frontier model, stochastic cost frontier model, and stochastic distance function model. Before selecting a specific model, analysts have to make an initial choice between the two most widely used functional forms: Cobb-Douglas function and translog function.

Cobb-Douglas: An advantage of Cobb-Douglas functional form is that it requires the estimation of very few parameters (for instance, four coefficients in a Cobb-Douglas production function with three inputs). Furthermore, the coefficients are very easy to interpret. However, the Cobb-Douglas model assumes that all firms have the same production elasticities, the same scale elasticities, and unitary elasticities of substitution; these assumptions are quite strong.
Translog: By comparison, the translog functional form is the most commonly used flexible functional form. While it requires the estimation of many more parameters than the Cobb-Douglas (for instance, fourteen in a translog production function with three inputs and time trend variable), it does not impose the restrictions imposed by the Cobb-Douglas. Another advantage of using the translog function is that the practitioner can decompose the TFP change into the technical efficiency change, technology change, and scale efficiency change. Therefore, academics generally prefer to use the translog functional form. However, the translog function requires a relatively large sample size and more complicated estimation process, such as imposing the assumption that the function is homogeneous of degree one in input prices and/or joint estimation of first-order cost share functions.

Therefore, for analysts with very limited sample size and experience, it is better to *start* with the Cobb-Douglas functional form. After accumulating the data for additional years and gaining experience over time, the regulator can consider utilizing the translog functional form.

Stochastic Production Frontier Model

The Stochastic Production Frontier describes output (e.g., water delivered) as a function of inputs (e.g., capital, labor, energy). The difference between the actual output and estimated (predicted) output is attributed to inefficiency and random noises.

Data requirements: quantity data on inputs and outputs for cross sectional firms or panel sample. The panel sample does not need to be balanced.
Advantages: no need for price information; efficiency change can be decomposed into technical efficiency change, technology change, and scale efficiency change.
Disadvantages: only one output can be taken into account; there is a potential endogeneity problem because input factors in a production function might be jointly determined with the output produced.

Stochastic Cost Frontier Model

Stochastic Production Frontier describes cost as a function of input prices and outputs. Difference between the real cost and estimated cost is attributed to inefficiency and random noises.

Data requirements: quantity data on inputs and outputs and price data on inputs for cross sectional firms or panel sample; the panel sample does not need to be balanced.
Advantages: can take multiple outputs into accounts; no potential endogeneity problem; efficiency change can be decomposed into cost efficiency change, technology change, and scale efficiency change; can calculate the allocative efficiency change if the results of the Tornqvist index are combined.
Disadvantages: might be difficult to get accurate input prices.

Stochastic Frontier Distance Function Model

A distance function may have either an input or an output orientation. An input orientation looks at how much the input vector may be proportionally contracted, holding the output vector fixed; an output orientation looks at how much the output vector may be proportionally expanded with the input vector held fixed. The distance function can be thought of as a multiple output version of a production frontier.

Data requirements: quantity data on inputs and outputs for cross sectional firms or panel sample; the panel sample does not need to be balanced.
Advantages: takes into account multiple outputs and inputs simultaneously; no need for price information; no assumption about the cost minimization; avoids the

potential endogeneity problem in production function; efficiency change can be decomposed into technical efficiency change, technology change, and scale efficiency change.
Disadvantages: relatively strong assumptions of the distance functional form, difficult to interpret the coefficients.

Corrected Ordinary Least Squares

Similarly, the COLS approach includes production function and cost function options.

Data requirements: quantity data on inputs and outputs for production function; quantity data on inputs and outputs and price data on outputs for cost function.
Advantages: easy to calculate, estimation is based on Ordinary Least Square (OLS); results are generally close to SFA models; can be used for robustness checks.
Disadvantages: assumes all the deviations are due to inefficiency.

Fixed or Random Effects Models (Distribution Free Methods)

The most widely used distribution free models are fixed effect and random effect models. These two models do not impose specific distribution assumptions onto the inefficiency terms, whereas in the stochastic frontier models the inefficiency term is assumed to follow specific distributions such as exponential, half normal and truncated normal distributions. However, each of these methods has its own limitations.

Fixed effects model: Assume the inefficiency term is captured by the time-invariant inefficiency term in fixed effect model. Firm-specific intercepts are estimated and the exponential term of the difference between the firm-specific intercept and the largest intercept among all the firms is regarded as the firm specific efficiency score.

Data requirements: quantity data on inputs and outputs for production function; quantity data on inputs and outputs and price data on outputs for cost function.
Advantages: no distribution assumption of the inefficiency term; easy to estimate.
Disadvantages: imposes strong assumption that the inefficiency is time invariant; the fixed effect (time invariant unobserved term) captures not only the inefficiency but also the effects of all phenomena (such as regulatory environment, ownership) that vary across utilizes but that are time invariant to each utility.

Random effects model: Random effects models relax the assumption that inefficiency is time invariant. In the random effect model, inefficiency terms are assumed to be randomly distributed with constant mean and variance, but are also assumed to be uncorrelated with the other regressors (in fixed effect models, inefficiency terms are allowed to be correlated with the other regressors). Firm-specific efficiency is estimated as the exponential form of difference between the average firm-specific residuals over time and the largest average firm-specific residual among the sample.

Data requirements: quantity data on inputs and outputs for production function; quantity data on inputs and outputs and price data on outputs for cost function.
Advantages: no distribution assumption of the inefficiency term, allows the presence of time-invariant regressors.
Disadvantages: imposes the strong assumption that the inefficiency term is not correlated with the other regressors; the estimation may be biased if the time period is very short.

These nine analytic techniques (identified in Figure 5) illustrate the range of methodologies available to those examining the determinants of utility output and utility costs. Since the resulting performance scores are only as accurate as the underlying models, the application of these approaches requires substantial technical skill. The key lesson for practitioners is that each method has strengths and limitations.

Appendix 4 Benchmarking in Regions of the World

This Appendix provides a brief overview of benchmarking initiatives in different regions of the world. It is not a comprehensive census, but is meant to illustrate the range of activities that characterize developing and developed nations. Furthermore, it does *not* try to benchmark the benchmarking activities in the different regions! Particular individuals or organizations contribute to initiatives, which sometimes do not take root, and in other cases evolve into collaborative projects. In any case, performance benchmarking of water supply utilities is now a common practice worldwide. As has been noted, the results of benchmarking programs provide ministries, agencies, and governments with an overview of current water supply status, as well as tools for improving utility performance. Furthermore, benchmarking can be an important planning tool for utility management.

In 2001, an International Benchmarking Review by WRc (international consultancy in water and the environment) identified about 160 benchmarking initiatives associated with about 700 water and sanitation utilities in over 100 nations. About one-third of the projects emphasized process benchmarking, although the activity in metric benchmarking focused on Key Performance Indicators rather than quantitative estimates of cost and production models. Many of these initiatives are beginning to have an impact on data availability, which is essential for the analysis of water utility performance. A catalogue of groups would include the World Bank Water and Sanitation Program, the International Water Association, Global Water Partnership (integrated water resource management–IWRM), the International Water Management Institute (especially in the water resource area), and many other organizations. The growth in attendance at World Water Forums (every three years) is a testimony to global interest in access to clean and safe water—participants come from international organizations, NGOs, national governments and other groups. One could build a case that such events tend to be heavy on rhetoric and light on quantifying sector performance. Promises and projections are easier to make than critical appraisals of actual accomplishments. Nevertheless, international, regional, and national meetings provide opportunities for networking and sharing best practice.

The short overviews presented below describe organizations and initiatives that are making metric benchmarking possible.

Latin America: ADERASA[1]

During the last fifteen years, most countries in the Latin American region introduced regulatory frameworks for the water and sewerage sector and created agencies to oversee and enforce them. Reforms were introduced in Chile and Argentina in the early 1990s, spreading to Bolivia, Colombia, and Peru in the mid 1990s, and to Nicaragua, Panama, Paraguay, Uruguay, and Venezuela towards the close of the decade. In Brazil and Mexico, systematic adoption of regulations has been limited to a few key states, while in Costa Rica a long-standing regulatory tradition was already in place. As a result, it is estimated that by the year 2000, about 40% of urban water consumers in Latin America enjoyed some form of regulatory protection.

Regulators often find themselves trying to apply modern regulatory techniques to a highly atomized and politicized sector of old-style municipal utilities. Benchmarking has proven to be an effective way to reduce information asymmetries and to promote public awareness of water utility performance. In the early stages of this process, there was a strong reliance on north-south transfer, with the concomitant problems of translating regulatory models across very different political and institutional environments. In October 2001, representatives of 10 Latin American regulatory entities (including Argentina, Bolivia, Chile, Colombia, Costa Rica, Nicaragua, Panama, and Peru, plus observers from the Dominican Republic and Venezuela) convened in Cartagena, Colombia to form a regional association of water regulators: ADERASA (Asociación de Entes Reguladores de Agua y Saneamiento de las Américas, or Association of Water and Sanitation Regulatory Entities of the Americas). The objectives of ADERASA were to promote cooperation and coordination of efforts in the development of the water sector in Latin America by facilitating the exchange of experiences and collaboration around common initiatives in the field of regulation.

[1] The primary author is grateful to Maria Luisa Corton and Alejo Molinari for providing background information for this section.

Its second annual meeting took place in Santa Cruz, Bolivia in September 2002, and was attended by approximately 100 staff from member countries, and regulatory body representatives from Brazil,[2] Honduras, Paraguay, and Venezuela, countries that subsequently joined the association. Thus, ADERASA brought together countries at varying stages of their development of a regulatory framework (e.g., Chile and Argentina with at least 10 years of experience, and Nicaragua and Venezuela with recently-created regulatory entities). This created opportunities for south-to-south exchanges, which can be more fruitful than traditional north-to-south exchanges. The development of an accurate and consistent regional database on benchmarking parameters was a priority for the organization. A core group of Argentinean members developed Performance Indicators (PIs) based on their own internal benchmarking experiences and other experiences in the region. The proposal included about 80 Performance Indicators, covering all relevant aspects of water and sewerage services. The definitions assigned to these PIs reflect the definitions used by the World Bank (created in 1999 as the Benchmarking Start-Up Kit, eventually becoming the International Benchmarking Network [IBNET]), and the International Water Association (IWA), paving the way for future global comparisons.

In 2003, a Task Force discussed a proposed manual with the assistance of the World Bank and IWA. That discussion resulted in a significant reduction in the number of proposed PIs, as well as modifications in some definitions. Later meetings focused on verification issues, and ultimately led to the first econometric analysis using data from 2003. In 2009, an updated statistical study conducted by consultants under ADERASA auspices utilized data from 2003 to 2008.[3] This group's work is probably the most technical (compared to other regional associations) though its application to actual rate cases is unclear. Data on national performance indicators tends to be utilized by regulatory commissions, though we can expect greater use of cross-country comparisons in the future as data consistency issues are resolved.

This regional collaboration illustrates the value of coordinated data collection initiatives. In addition, the Inter-American Development Bank (IADB) and

---

[2] In the case of Brazil, the National Sanitation Information System (SNIS; http://www.snis.gov.br/oque_snis.htm) is a federal data base started in 1995 and maintained by the City Ministry of Brazil (MCIDADES), under the direction of its National Secretary for Environmental Sanitation (SNSA). More recently, the Association of Brazilian Regulators (ABAR–Associação Brasileira de Agências de Regulação) has taken on the task of capacity building and the promotion of benchmarking activities.

[3] Gustavo Ferro and Carlos Adrián Romero (2009). "Estudio de Fronteras de Eficiencia: Empresas de agua y alcantarillado de América Latina, ADERASA, September, 1–105.

UN-Habitat have initiated a Water Operators Partnership promoting cooperation in many areas: "institutional, legal, managerial, related corporate governance, transparency, accountability, financing, commercialization," operation, and investment. Although a country-by-country overview is beyond the scope of this volume, much more information about the regional and national initiatives is available on the ADERASA website (which includes links to national water regulators). Many multinational organizations attempt to address water sector issues. For example, the Central American water and sanitation network (Red Regional de Agua y Saneamiento de Centroamérica, RRAS-CA, founded in 1992) facilitates information sharing—more in the area of health and sanitation than on water sector performance. The IADB recently funded a benchmarking study of Central American water utilities, suggesting that there could be greater emphasis on performance comparisons than in the past.[4]

Africa: WUP

The Water Utility Partnership (WUP) for Capacity Building is a partnership organization founded by the Union of African Water Suppliers (UAWS) and the International Training Network Centers in Africa – CREPA and TREND. After a 1996 meeting on institutional reform in the Water Sector, WUP was established to address key challenges facing utilities and their partners in the region. The primary constituency for WUP is the Union of African Water Suppliers (UAWS), a membership association of utilities in Africa. Another organization, the South African Association of Water Utilities (SAAWU) also sought to promote the efficiency and financial sustainability of water utilities. It launched a benchmarking project in 2001, focusing on forty-nine Key Performance Indicators.

WUP initiated a parallel project, involving utilities from throughout Africa. SPBNET—Africa 2001 was led by WUP and overseen by a Quality Assurance Committee consisting of WUP, the World Bank, a utility regulator, the Union of African Water Suppliers, International Water Association, and the UK's Department for International Development (DFID). It received technical advice from the UK's Water Research Centre (WRc). Ultimately, 110 utilities from 40 African countries provided performance information. It is somewhat surprising that this initiative did not lead to a comprehensive database that is being updated on an annual basis. The efforts have been somewhat fragmented.

[4] Maria Corton and Sanford Berg (2009). "Benchmarking Central American Water Utilities," *Utilities Policy*, **17**(3–4), 267–275.

The African Forum for Utility Regulators (AFUR), established in 2002, represents another potential catalyst for data collection and analysis—in this case from the side of those agencies providing oversight. Their Water and Sanitation Sectoral Committee would be the logical group to collate and make comparative data available. In addition, a number of regulatory commissions are collecting and publishing comparative performance data from utilities. For example, Zambia's National Water Supply and Sanitation Council (NWASCO) established benchmarks for six standard core indictors (like labor productivity and collection efficiency). On the operator side, Uganda's National Water and Sewerage Company (NWSC) utilizes econometric models to set performance targets and incentives for managerial compensation.

Currently, the African Water Association in cooperation with the World Bank Institute and the NWSC (Uganda) are planning to establish an African Water Academy to promote capacity building and improved management in the water and sanitation sector. Thus, we observe a great deal of activity in the region, though systematic collection and analysis of data lags. The extant cost and production function studies have tended to be published by researchers at the World Bank.[5]

Asia: SEAWUN[6]

The South East Asia Water Utility Network (SEAWUN), established in 2002 in Hanoi is another regional organization designed "to help the members improve their performance in the delivery of water supply and sanitation services for all, including operation and management efficiency, achieving financial viability, and advocating for sector reforms for improved policy environment, contributing to realize the goal 'Water for All.'" SEAWUN's current membership includes five water associations, twenty-five water related companies from Indonesia, Vietnam, Thailand, Malaysia, and the Philippines, and one consulting company in Australia. In its early stages, financial support was provided by Asian Development Bank (ADB) and others from within the SEAWUN membership. The general purpose of ADB's support is to substantially improve the delivery of water supply and sanitation (WSS) services in South East Asian developing-member countries, including the poor.

---

[5] A. Estache and Kouassi, E. (2002). "Sector organization, governance, and the inefficiency of African water utilities," World Bank, Research Working Paper No. 2890.

[6] The author is grateful to Liangliang Jiang for assembling some of this information.

Although of great importance and use, performance benchmarking has not yet been intensively implemented in many countries in the region. Utilities in smaller South East Asian countries have little internal support for performance benchmarking. The scope of the performance benchmarking program includes:

- review of previous and ongoing benchmarking programs,
- development of a basic set of performance indicators (PIs).

The initial benchmarking survey (2004–05) obtained data from 47 water utilities in seven countries (Indonesia, Malaysia, Thailand, Philippines, Vietnam, Laos, and Cambodia). Initially, the survey only collected raw data (161 data fields for 61 questions in 10 categories). Then, a variety of performance indicators were calculated (82 indicators for 13 areas in 4 subjects: Customer Indicators, Water Supply System Operations and Management, Human Resources, and Finance). To obtain an overall comparison, 14 of these indicators were used to calculate an overall performance indicator (OPI). Unlike the IBNET survey, the SEAWUN BM survey focused only on water supply.

The criteria for developing a performance analysis system are encapsulated in the acronym FUEL—flexible, useful, easy, and level. Compared with other surveys, the SEAWUN data collection requirements are relatively modest: 61 questions, compared to about 92 water-supply related questions in the IBNET survey. On the other hand, it included subjects such as the separation of unaccounted for water (UFW), as recommended by the International Water Association (IWA). In addition, it included questions on investment in human resources and infrastructure.

The Asian Development Bank (ADB) has expanded its support of benchmarking to the South Asian Water Utilities Network (SAWUN), created in 2007. Like SEAWUN, the organization promotes capacity-building and the sharing of best-practice principles. The region includes Bangladesh, Bhutan, India, Maldives, Nepal, and Sri Lanka. In addition, the World Bank, the Private Public Infrastructure Advisory Facility (PPIAF), and AusAID (Australia) have funded the East Asian and Pacific Infrastructure Regulatory Forum (EAPIRF), created in 2003. With fifty member organizations from twenty-four nations, the organization promotes information-sharing and capacity-building in the region. It is expected that the association of regulators and the association of suppliers will interact on a more regular basis—leading to more studies using the metric benchmarking techniques described in this volume. Similarly, the South Asia Forum for Infrastructure Regulation (SAFIR), established in 1999, is providing a venue for capacity building and data collection in that region.

OECD[7]

Water utilities in wealthier nations have engaged in process benchmarking for some time. Performance comparisons are also sponsored by some national regulators. Ofwat, the water regulator for England and Wales, has promoted statistical benchmarking studies for nearly two decades. The IWA volumes document the types of studies conducted under the auspices of water utilities and various government agencies. To consolidate some of these efforts, the European Benchmarking Co-operation (EBC) brought together the national water associations of Denmark, Finland, Norway, and the Netherlands in 2004. The focus goes beyond costs, to environmental sustainability, water quality, and finance. However, it has not emphasized metric benchmarking. The Portuguese water regulator has emphasized the kinds of metric benchmarking (efficiency models) in its evaluation of sector performance. The American Water Works Association published a volume on *Selection and Definition of Performance Indicators for Water and Wastewater Utilities* in 2004, focusing on 22 performance indicators. In conjunction with the IWA Specialist Group on Efficient Operations and Management, an EU initiative on performance benchmarking led to a conference and book describing many projects.[8] All these initiatives demonstrate the interest in the field, although the application of metric benchmarking still tends to be done more by academics than those developing and implementing policy.

In Scandinavia, water supply and wastewater services are normally provided by the local authorities, which both operate the systems and own the assets. In small communities in Denmark, Finland and Norway, the water supply is often managed by consumer cooperatives, while the municipality provides sewerage. In order to increase cooperation and improve efficiency in water and wastewater utilities, the Scandinavian 6-cities Group – Copenhagen, Oslo, Helsinki, Stockholm, Gothenburg and Malmo – developed a number of performance indicators and started cooperating in the area of benchmarking in 1995. The focus tends to be on Performance Indicators rather than on economic models of cost and production. Water utilities track their system performance over time and in comparison to others in the group. Meetings focus on process benchmarking, to increase efficiency by "learning from others".

---

[7] The author is grateful to Liangliang Jiang for preparing material utilized in this section.

[8] Enrique Cabrera, Jr. and Miguel Ángel Pardo (2008). *Performance Assessment of Urban Infrastructure Services*, IWA Publishing, 515–523.

A project on benchmarking methodologies for the Australian Water Industry was sponsored by the Water Authority of Western Australia and funded by the Urban Water Research Association of Australia. The major objectives are the development and pilot testing of a benchmarking methodology suitable for use in the Water Industry. The secondary objective is to train a core of Water Industry personnel in the application of this methodology. Core indicators in benchmarking water and sanitation utilities in Australia are contained in two different datasets: from the Water Services Association of Australia (WSAA), consisting of performance indicators for the urban water industry from 1997 to 2004, and from the Economic Regulation Authority, consisting of water performance information on 32 major Western Australian towns from 1999 to 2003.

The *Canadian National Water & Wastewater Benchmarking Initiative* was launched in 1997 and has since grown to 38 member municipalities participating in storm water, water and wastewater benchmarking. The Benchmarking Initiative was established to translate the results of metric benchmarking into actions that are driving performance improvement. By visiting the participating utilities instead of using questionnaires, a number of problems have been avoided, where typical problems include inconsistent data, incomplete questionnaires or the non-return of questionnaires. Also, all data input into the database were confirmed through a reference to the utility's financial records, annual report or other written records. It is interesting to note that cost or production function models have not been estimated using the dataset.

Thus, OECD nations have a number of initiatives. Each country or region has its own unique regulatory or governance system—making cross-country collaborations somewhat difficult, especially for the larger nations which have their own information-gathering systems. Nevertheless, a general pattern is emerging: groups responsible for developing and implementing public policy and managers of water utilities recognize the value of performance comparisons—as a way to identify best practice and as a stimulus to improved operations.

Index

IWA Publishing's authorised EU representative for General Product Safety Regulations is Diane D'Arras, 15 rue Duret, 75116 Paris, France, e-mail: safety@iwap.co.uk.

Printed and bound by CPI Group (UK) Ltd, Croydon, CR0 4YY
16/06/2026
02140405-0001